Bibliographic information published by the German National Library:

The German National Library lists this publication in the National Bibliography; detailed bibliographic data are available on the Internet at http://dnb.dnb.de .

Imprint:

Copyright © 2016 GRIN Verlag, Open Publishing GmbH
Print and binding: Books on Demand GmbH, Norderstedt Germany
ISBN: 9783668375482

This book at GRIN:

http://www.grin.com/en/e-book/350021/near-fault-seismic-site-response-through-observed-and-simulated-data-of

Ermanno Ragozzino

Near-fault seismic site response through observed and simulated data of the 2009 L'Aquila (central Italy) Mw 6.1 earthquake

GRIN Publishing

GRIN - Your knowledge has value

Since its foundation in 1998, GRIN has specialized in publishing academic texts by students, college teachers and other academics as e-book and printed book. The website www.grin.com is an ideal platform for presenting term papers, final papers, scientific essays, dissertations and specialist books.

Visit us on the internet:

http://www.grin.com/

http://www.facebook.com/grincom

http://www.twitter.com/grin_com

To my father Giuseppe and my son Joseph.

Title

Near-fault seismic site response through observed and simulated data of 2009 L'Aquila (central Italy) M_w 6.1 earthquake

Author name and affiliations

Name: Ermanno Ragozzino
Affiliations: Lazio Region, Regional Directorate for Infrastructure and Housing Policy, Roma, Italy
E-mail: eragozzino@regione.lazio.it

Corresponding author

Ermanno Ragozzino
E-mail: eragozzino@regione.lazio.it

Contents

Introduction

During the night of 6 April 2009, at 01:32 GMT, an earthquake hit the L'Aquila Basin, in central Italy. The earthquake caused damage to between 3000 and 11000 buildings in the medieval center. Several buildings also collapsed. Three hundred and nine people died and more than 1500 were injured. About 65000 people (out of a population of 72000) had to abandon their homes. The Mw 6.1 main shock occurred with an epicenter at 42.3400° N, 13.3800 °E, approximately 90 km NE of Rome, near the city of L'Aquila. The earthquake was caused by movement on a NW-SE trending normal fault that was defined to strike 140° and dip 50° to the SW. Most co-seismic slip occurred in a small rupture area of approximately 16 km along strike and nearly 8 km in the dip direction, in a depth range of 4-10 km [41]. The main shock and many aftershocks of the 2009 L'Aquila seismic sequence were recorded by the near-fault stations Colle Grilli (AQG), Fiume Aterno (AQA), Centro Valle (AQV), Il Moro (AQM), Aquil Park Int. (AQK) that belong to the Italian Strong Motion Network (RAN) [20,51] and Aquila Castello (AQU) that belong to the National Institute of Geophysics and Volcanology (INGV). The stations are situated in the Upper Aterno River Valley and in the medieval City of L'Aquila. Later, several groups and institutions made a great quantity of data available as a result of geological, geotechnical and geophysical investigations [2,17,46], whereas various authors performed different types of research studies.

In particular, De Luca et al. [10] found evidence of low-frequency amplification at 0.6 Hz in the City of L'Aquila through analysis of earthquake and ambient noise data. They also performed 2D numerical modeling by means of both finite and boundary element methods, which allowed them to relate the low-frequency amplification to the presence of a sedimentary basin about 250 m deep.

Puglia et al. [34] performed spectral ratio analyses for the permanent seismic stations of the Upper Aterno River Valley array, which were based on earthquake and noise recordings and 1D elastic equivalent-linear modeling. They demonstrated a fundamental frequency shift from 3 Hz to 1.5 Hz at seismic station AQV during the 2009 M_w 6.1 main shock and suggested that one possible explanation was the nonlinear behavior of soil.

Chioccarelli and Iervolino [8] proposed that frequency change with time is not directly related to nonlinear shear modulus behavior, but to peculiar phases of the seismic signal whose frequency is more related to source effect such as directivity and flings.

Lanzo and Pagliaroli [26] estimated the seismic site effects in the Aterno River Valley through standard spectral ratio analysis of near-fault strong motion records by using seismic station AQG as reference site. They also performed 1D numerical modeling, but did not clarify the observed frequency change with magnitude level at seismic station AQV.

Del Monaco et al. [13] carried out microtremor recordings in historical Downtown L'Aquila and applied the Nakamura method [33], which revealed maximum amplification at the frequency ranges of 0.4-0.7 Hz and 3-15 Hz.

Gaudiosi et al. [19] focused their attention on a cross-section of the Upper Aterno River Valley, analyzed the strong and weak motion data by means of SSR (standard spectral ratio) technique and compared the results with those obtained from 2D numerical modeling that used both finite element and finite difference methods. They found evidence of a strong amplification at seismic station AQV, which was related to the constructive interference of S and surface waves. They also noticed nonlinearity in soil behavior.

Nunziata and Costanzo [33] applied a hybrid method (modal summation plus finite difference) to poorly defined geologic models of the Upper Aterno River Valley and Downtown L'Aquila and validated the numerical results with recordings at stations AQG, AQA, AQV, AQK and AQU. It resulted that: the alluvial sediments filling the Upper Aterno River Valley caused amplification at the frequency range of 2 and 7 Hz; L'Aquila breccia and sand, which outcrop in historical L'Aquila City, caused response spectral amplification at the frequency range of 0.6 and 7 Hz.

Considering the results of mentioned surveys, in this book I will try to present the outcomes of my research (36, 37), which is about the seismic response of sediments that fill L'Aquila Basin, by using two case studies: 1) Western L'Aquila Basin; 2) Historical L'Aquila City center. In particular, I will analyze the April 2009 recorded ground motions and compare the observed data with those revealed from finite element nonlinear dynamic models.

Case Study #1: Western L'Aquila Basin

The Aterno River Valley Strong-Motion Array

The strong-motion array in the upper Aterno River Valley (Figure 1), which is situated in Western L'Aquila Basin, is a part of the Italian Strong Motion Network. It was installed in 1994 to investigate the seismic site effects along a line transversal to the valley [4]. The array originally consisted of seven stations and was modified over the years. It is currently composed of six digital strong-motion instruments, which are named Colle Grilli (AQG), Fiume Aterno (AQA), Centro Valle (AQV), Il Moro (AQM), Ferriera (AQF) and Monte Pettino (AQP).

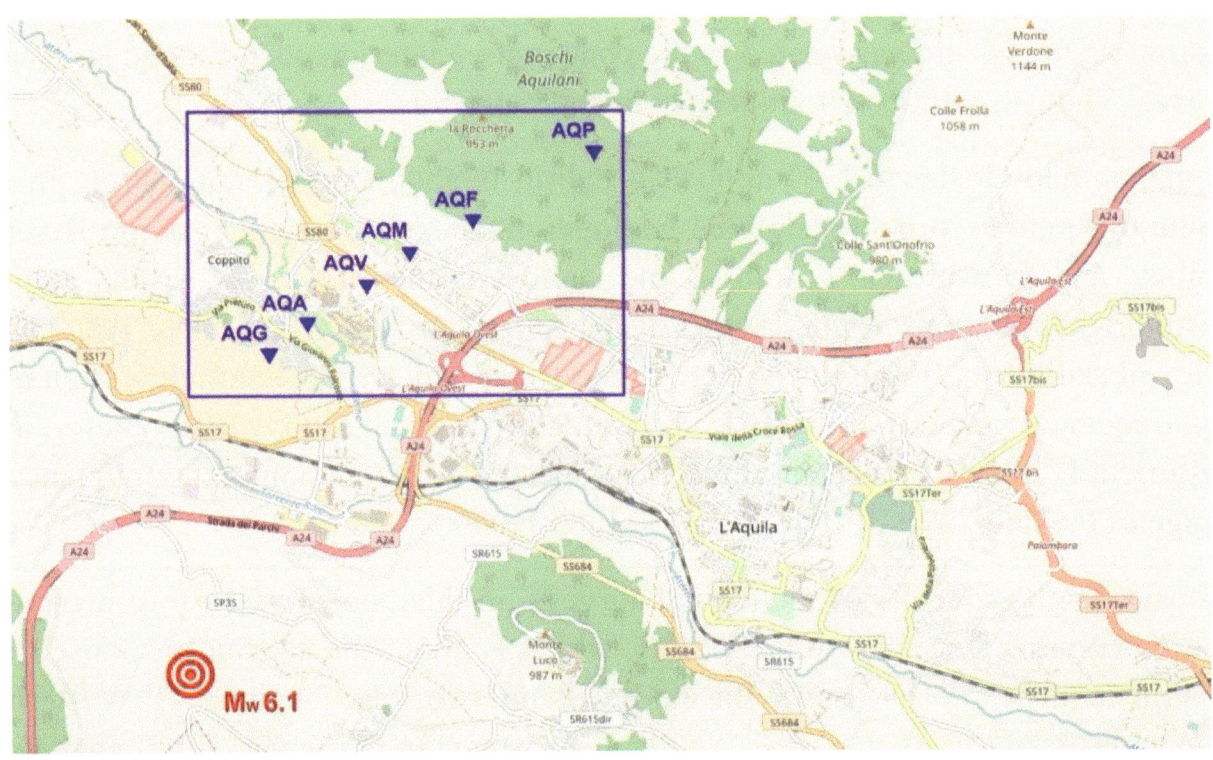

Fig. 1. Study area map showing the Aterno River Valley strong-motion array and the epicenter of 2009 M_w 6.1 main shock.
Map © OpenStreetMap contributors, http://www.openstreetmap.org/copyright/en

These stations are notably near the hanging wall of Mt. Pettino master extensional fault [15]. The array is approximately 2 km long, with a 250 m station spacing. Stations AQG, AQM and AQP are placed on Meso-Cenozoic carbonate rock outcrops; AQA and AQV are on the Aterno River alluvial deposits; and AQF is on the Mt. Pettino debris flow and pediment alluvial deposits. With the exception of AQF, which recorded only one event of the 2009 seismic sequence, the stations and the records that are considered for the analyses and subsequently used as the input motion in the 2D dynamic model, are summarized in Table 1. Stations AQF and AQP did not record the main shock because of a power supply failure. Station AQM recorded the main shock and was affected by saturation (PGA = 1 g) [38], but the data were not formally released because the recorded motion could be disturbed by earthquake-induced damage to the seismometer protective box [3].

Table 1
Analyzed events (recorded in the period 1998-2009) and maximum acceleration (PGA) according to the magnitude levels (M_L and M_W) for the examined seismic stations.
M_L is the– local magnitude and M_W is the– moment magnitude.

| Station | M_L 1.6-3.7 | | M_W 4.1-5.6 | | M_W 6.1 | |
	Events	PGA(cm/s^2)	Events	PGA(cm/s^2)	Events	PGA(cm/s^2)
AQG	13	0.06-8.35	9	18.8-133.2	1	479.3
AQA	13	0.12-6.46	5	15.6-56.8	1	435.4
AQV	17	0.14-12.9	9	30.2-338.8	1	644.2
AQM	13	1.5-6.5	8	10.8-231.7		
AQP	7	1.3-4.3	7	19.3-92.1		

Geological, Geotechnical and Geophysical Data

The geologic setting of Western L'Aquila Basin (Figures 2 and 3) consists of a Quaternary sediment-filled alluvial valley, which lies over a Meso-Cenozoic carbonate bedrock [15,45,47]. The Quaternary deposits are mostly represented by granular soils and reach a maximum thickness of approximately 50 meters near the station AQV. The carbonate bedrock, which includes flinty limestone (Corniola Formation) and intercalations of reef-slope facies detrital limestone (Maiolica Formation, Calcarenites and calcareous breccias with Fucoids), is structured as an asymmetric graben. It is variously dislocated by a series of northwest-southeast and northeast-southwest normal faults and a back-thrust; the latter brings the Maiolica Formation over the Miocene arenaceous-pelitic turbidites. Geological and geophysical data, which are obtained from the borehole logs and in situ seismic tests (Figures 4 and 5) [4,16,25,44,46], allow us to subdivide the investigated lithotypes into 10 geotechnical units and define the 2D subsurface model along the Aterno River Valley strong-motion array (Figure 3). Among the Quaternary deposits, it is possible to recognize seven geotechnical units (A, B1, B2, C, D1, D2 and E in Figures 3, 4 and 5) and three units (F1, F2 and F3 in Figures 3, 4 and 5) in the Meso-Cenozoic carbonate bedrock according to the soil and rock geotechnical properties and seismic wave velocity. Boreholes S4, S3, S5 and S1 also helped to define the water table in the alluvial deposits at 3.8, 5.1, 5.3 and 14.8 meters under the ground level, respectively (Figure 4). The geotechnical units are summarized in Table 2 with the geotechnical and geophysical features that are obtained from laboratory and in situ tests.

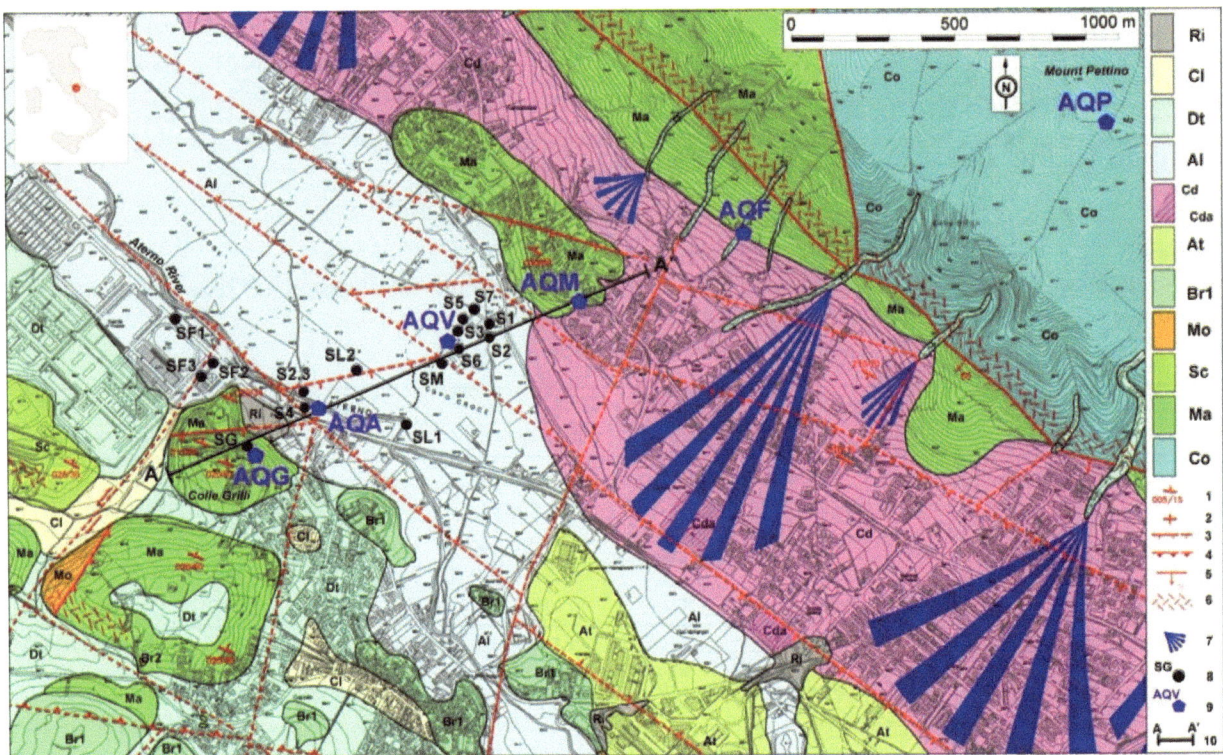

Fig. 2. Geologic map of the studied area and the surrounding areas. Geological units. Quaternary units: Ri – anthropogenic fill material (Holocene); Cl – colluvium: fine-grained deposits (Holocene); Dt – debris slope deposits (Holocene); Al – Aterno River alluvium (Holocene); Cd – Mt. Pettino debris flow and pediment alluvial deposits: dense and poor to well-cemented calcareous gravel with tephra horizons; Cda – micro-karst calcareous gravel (upper-middle Pleistocene); At – Vetoio Stream terraced alluvium: gravel, sand and clayey-sandy silt with tephra horizons (upper-middle Pleistocene); Br – L'Aquila breccia: dense and poor to well-cemented calcareous gravel (middle Pleistocene). Meso-Cenozoic carbonate units: Sc – detrital Scaglia Formation: mudstone and fine grainstone (Eocene -Cenomanian); Ma – detrital Maiolica Formation, Calcarenite and calcareous breccias with fucoids: mudstone, fine grainstone and oolitic limestone (Cenomanian-upper Tithonian); Co – detrital Corniola Formation: mudstone and fine grainstone, which are sometimes dolomitized (middle Lias). Syn-orogenic unit: Mo – pelitic-arenaceous turbidites (middle Miocene). 1 – dip direction and angle; 2 – horizontal layers; 3 – fault; 4 – overthrust fault; 5 – fault dip; 6 – fault rock; 7 – alluvial fan; 8 – borehole; 9 – seismic station; 10 – geologic cross-section line. Modified after [47].

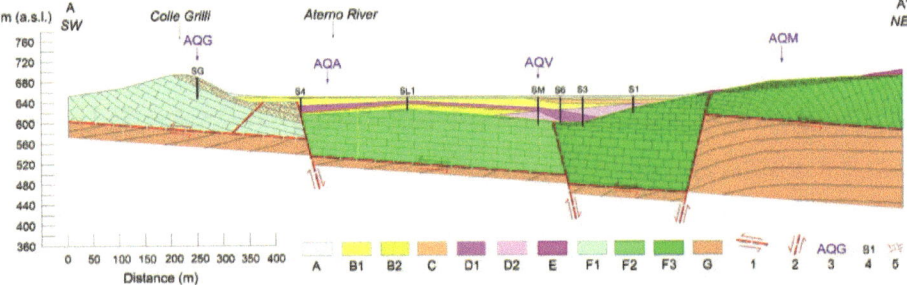

Fig. 3. Geologic cross section of the studied area (its line is positioned in Figure 2). Geotechnical units. Quaternary units: anthropogenic fill material and colluvium: A – gravelly silty clay, clayey silt with gravel; Aterno River alluvium: B1 – silty clayey gravel with sand, poorly graded gravel, sand and silt with gravel; some level of clay with gravel; C – sandy silty clay with gravel, sandy silt with gravel, clay with sand, silty clay, silty clayey gravel with sand; B2 – poorly graded silty clayey gravel, well graded coarse sand; some level of gravelly silty clay; D1 – sandy silt and clayey silt with sand and gravel; poorly graded silty clayey gravel with sand, silty sand, sandy silt; D2 – sandy silt, silt with clay and sand sometimes with gravel, sandy silty clay sometimes with

gravel, alternations of sandy silt-silty sand-silty gravel with sand; Mt. Pettino debris flow and pediment alluvial deposits: E – dense and poor to well-cemented calcareous gravel with tephra horizons. Meso-Cenozoic carbonate bedrock: F1 – highly fractured; F2 – moderately fractured; F3 – slightly fractured. Syn-orogenic unit: G – pelitic-arenaceous turbidites. 1 – back-thrust fault; 2 – normal fault; 3 – seismic station; 4 – borehole; 5 – fractured and weathered limestone.

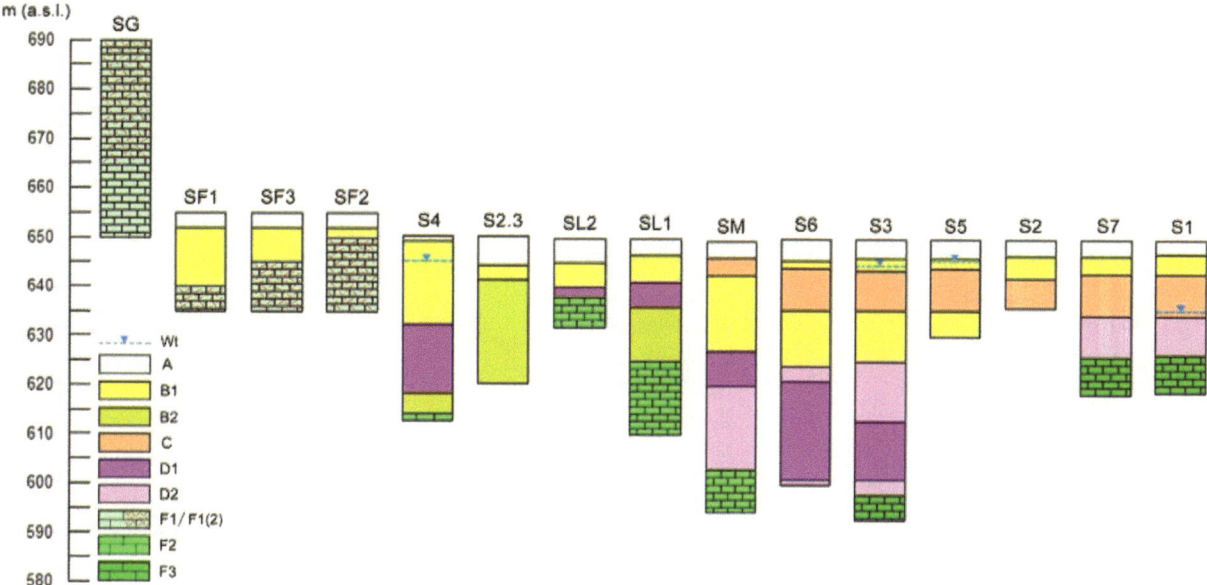

Fig. 4. Borehole logs of the investigated area (their position is in Figure 2). Geotechnical units. Quaternary units: anthropogenic fill material and colluvium: A – gravelly silty clay, clayey silt with gravel; Aterno River alluvium: B1 – silty clayey gravel with sand, poorly graded gravel, sand and silt with gravel; some level of clay with gravel; C – sandy silty clay with gravel, sandy silt with gravel, clay with sand, silty clay, silty clayey gravel with sand; B2 – poorly graded silty clayey gravel, well graded coarse sand; some level of gravelly silty clay; D1 – sandy silt and clayey silt with sand and gravel; poorly graded silty clayey gravel with sand, silty sand, sandy silt; D2 – sandy silt, silt with clay and sand sometimes with gravel, sandy silty clay sometimes with gravel, alternations of sandy silt-silty sand-silty gravel with sand. Meso-Cenozoic carbonate bedrock: F1 – highly fractured; F1(2) – highly fractured and weathered; F2 – moderately fractured; F3 – slightly fractured. Wt – water table.

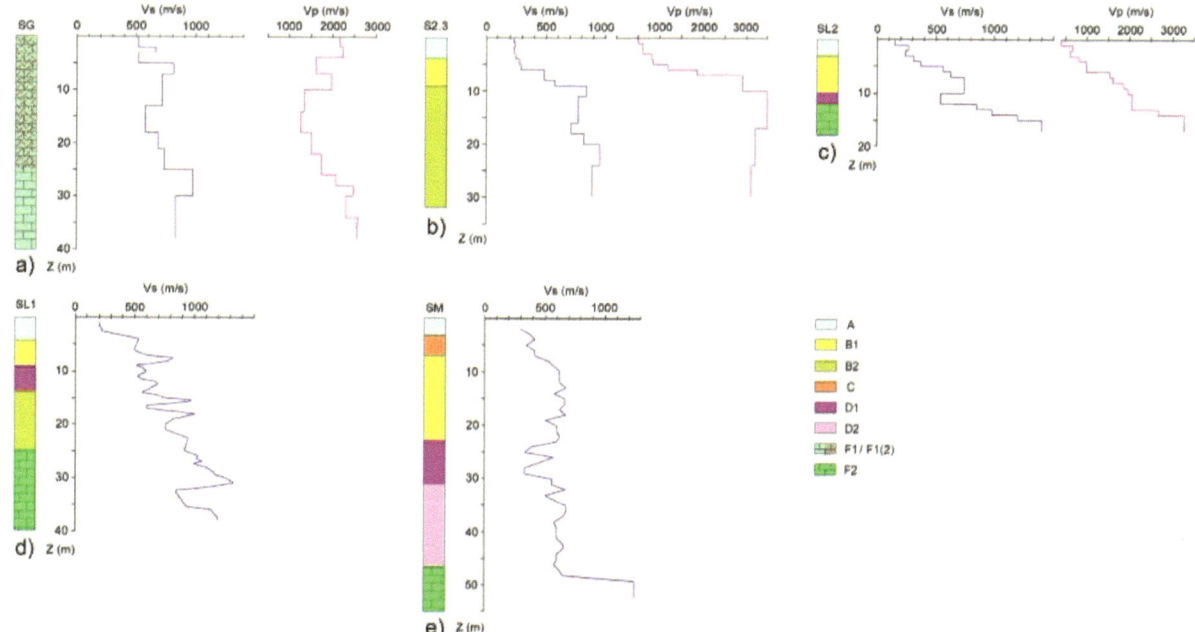

Fig. 5. Borehole logs with P and S wave velocity profiles (their position is in Figure 2). Geotechnical units. Quaternary units: anthropogenic fill material and colluvium: A – gravelly silty clay, clayey silt with gravel; Aterno River alluvium: B1 – silty clayey gravel with sand, poorly graded gravel, sand and silt with gravel; some

level of clay with gravel; C – sandy silty clay with gravel, sandy silt with gravel, clay with sand, silty clay, silty clayey gravel with sand; B2 – poorly graded silty clayey gravel, well graded coarse sand; some level of gravelly silty clay; D1 – sandy silt and clayey silt with sand and gravel; poorly graded silty clayey gravel with sand, silty sand, sandy silt; D2 – sandy silt, silt with clay and sand sometimes with gravel, sandy silty clay sometimes with gravel, alternation of sandy silt-silty sand-silty gravel with sand. Meso-Cenozoic carbonate bedrock: F1 – highly fractured; F1(2) – highly fractured and weathered; F2 – moderately fractured.

Table 2
Mean geotechnical and geophysical features of the units. γ – unit weight; c – cohesion; φ – friction angle; V_S – S-wave velocity; V_P – P-wave velocity.

Unit	Prevalent Lithotype	Fracturing/ weathering	Consistency	γ (kN/m³)	c (kPa)	φ	V_S (m/s²)	V_P (m/s²)
A	Silt and clay with gravel		Plastic	19	30	22°	250	900
B1	Gravel, sand, silt		Medium Dense	18.5	5	36°	650	1500
B2	Gravel, sand		Dense	19	5	45°	850	3000
C	Sandy silt, clay		Fluid/Plastic	18	33	23°	400	900
D1	Silt		Very Soft/Soft	17.5	35	23°	500	2000
D2	Silt		Dense/Plastic	19	30	22°	600	1500
E	Gravel		Dense/Very Dense	19	15	29°	550	1200
F1	Limestone	High		22	250	30°	850	1800
F2	Limestone	Moderate		23	300	35°	1200	2500
F3	Limestone	Low		24	350	40°	1400	2650

Ground Motion Record Analysis

The strong- and weak-motion data are analyzed for 312 signals, including (the three-component acceleration waveforms), which were recorded using seismometers at the stations AQG, AQA, AQV, AQM and AQP during the events with magnitude M_w ranging from 1.6 to 6.1, in the period 1998-2009 (Table 1). The corrected acceleration time-histories are downloaded from the Italian Accelerometric Archive web site [28,46], viewed and analyzed to compute the observed maximum acceleration and velocity (Table 1) and the 5 % damping elastic acceleration response spectra [9,32]. The PGA (Peak Ground Acceleration) amplification factor of the recorded 2009 events with a magnitude range of 4.1 to 6.1 is computed based on the ratio between the recorded PGA at the site and the reference site during the same event, assuming that AQG and AQP are the reference stations. Nonetheless, these rock sites are not free of local effects [1] and are considered the "reference" sites to highlight the PGA amplification at the two soil stations AQA and AQV. (We are well aware of the limitations that this assumption implies).

Finally, the single station H/V (Horizontal-to-Vertical response spectral ratio) (10,18,21,27,29,40,42,43) and V/H (Vertical-to-Horizontal response spectral ratio) methods are used to evaluate the site effects at stations AQG, AQA, AQV and AQM.

Dynamic Modeling

To simulate the seismic site response at the four stations AQG, AQA, AQV and AQM, 2D dynamic analyses are performed on three parts (Figures 6a, b, c) of the geologic cross-section displayed in Figure 3, using QUAKE/W [35], (which is a software based on finite element formulations with a direct integration scheme in the time domain). The mesh pattern of the models consists of quadrangular and triangular elements (with global sizes of 7.5 m (Figures 6a and b) and 5 m (Figure 6c), respectively). The initial static analyses are performed to determine the initial state of stress in the ground before starting the dynamic analyses. The initial pore-water pressure conditions are computed by specifying the water table, which is revealed from the boreholes in Figure 4.

The lower boundary where the movement is fixed in both the x and y directions is positioned: for the model in Figure 6a, (the lower boundary is on the interface) between the Meso-Cenozoic fractured carbonate bedrock and the Miocene turbidites (Figure 3); for the models in Figures 6b and c, (the lower boundary is on the contact) between the Quaternary deposits and the Meso-Cenozoic carbonate bedrock (Figure 3).

To simulate the observed non-linear effective stress behavior (6,11,14,24,48,49), 122 nonlinear dynamic analyses based on the MFS (Martin-Finn-Seed) pore-water pressure model (7,24,50) are executed for each model (Figures 6a, b, c). The MFS model is based on the concept that the pore-pressure, which is generated during undrained loading, is related to the volumetric strain that would occur for the same stress increment under drained loading according to the following expression:

$$\Delta u = E_r \, \Delta \varepsilon_{vd}$$

where E_r is known as the rebound modulus (Figure 7a) and $\Delta \varepsilon_{vd}$ is the incremental volumetric strain that would occur under drained loading conditions. The expression for the incremental volumetric strain, which was developed by Martin [24], is the following:

$$\Delta \varepsilon_{vd} = C_1 \left(\gamma - C_2 \, \varepsilon_{vd} \right) + \frac{C_3 \, \varepsilon_{vd}^2}{\gamma + C_4 \, \varepsilon_{vd}}$$

where γ is the dynamic shear strain amplitude in the cycle and ε_{vd} is the plastic or accumulated volumetric strain (Figure 7b). C1, C2, C3 and C4 are curve-fitting constants and in this sense are material properties.

The geotechnical unit properties that are used as input parameters in the analyses are presented in Table 3. The rebound modulus E_r [24] and the MFS pore-water pressure functions (7,50) that are estimated for different materials are shown in Figure 7.

Ten-second windows of the horizontal and vertical acceleration time-histories are recorded by AQG, AQM and AQP during the 1998-2009 events with M_w ranging between 1.6 and 6.1 (Table 1), and the time windows are scaled and used as the input motion records. In particular, during each analysis the models (Figures 6a, b and c) are simultaneously subjected to the horizontal and vertical accelerations of the earthquake. Finally, the computed acceleration response spectra are normalized to calculate the numerical H/V and V/H response spectral ratios.

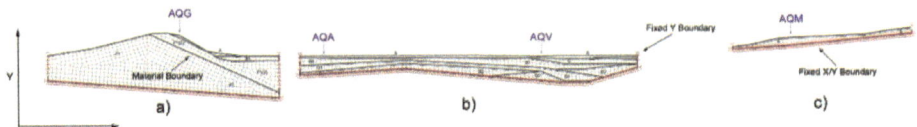

Fig. 6. Model setup of the material zones (A, B1, B2, C, D1, D2, E, F1, and F1(2) in Figures 3, 4 and 5), the boundary conditions and the mesh patterns that were used to perform the 2D dynamic modeling. AQG,…, AQM are the– seismic stations.

Table 3
Geotechnical and dynamic properties of the units used as input parameters in the numerical analyses.

Unit	A	B1	B2	C	D1	D2	E	F1	F1(2)
Unit weight (kN/m³)	19.0	18.5	19.0	18.3	17.5	19.2	19.0	24.0	23.0
Cohesion (kPa)	30	5	5	33	35	30	30	300	250
Friction angle (degrees)	22	36	45	23	21	22	28	35	30
Poisson's ratio	0.405	0.383	0.462	0.373	0.466	0.405	0.400	0.427	0.348
Damping ratio (min)	0.05	0.05	0.05	0.05	0.05	0.05	0.05	0.05	0.05
Damping ratio (max)	0.1	0.2	0.3	0.2	0.3	0.3	0.2	0.06	0.1
G_{max} (MPa)	112	800	1400	379	447	702	587	1967	1114

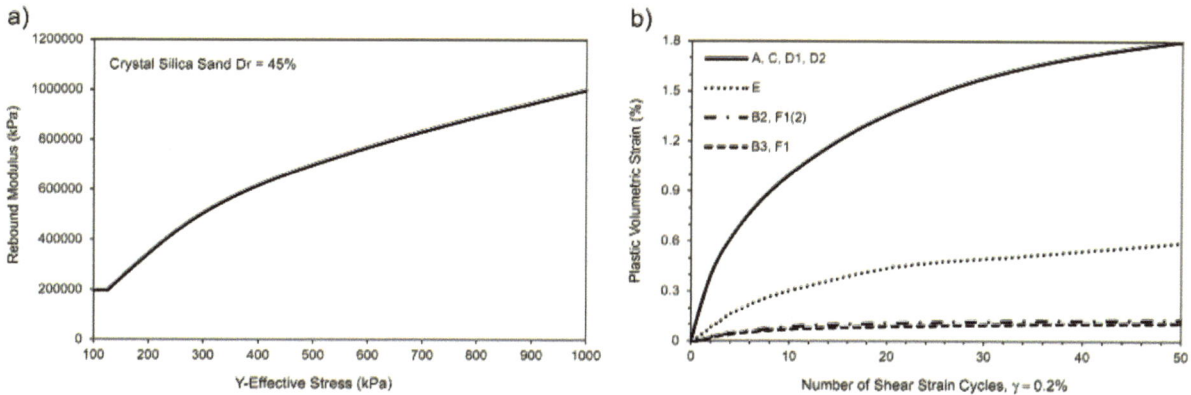

Fig. 7. a) Variation of the rebound modulus with the y-effective stress, which was obtained for silica sand at a relative density of 45% (Martin et al., 1975); b) Estimated variation of the plastic volumetric strain of the modeled materials with a number of shear strain cycles.

Peak Ground Acceleration Amplification Factor

For the recorded 2009 M_w 4.1-6.1 events, the horizontal (Figure 8a) and vertical (Figure 8b) PGA amplification factors display a maximum amplification at station AQV (on the thickest part of the alluvium), which can be related to the strong impedance ratios [22] that occur in the alluvium and between the alluvium and the carbonate bedrock (Figure 5e). The horizontal PGA amplification factor slightly decreases when the earthquake magnitude increases for stations AQG, AQA and AQV (Figures 9a, b, and c), which is related to the nonlinear behavior of the fractured-weathered rock (F1(2) in Figures 4 and 5a) and the alluvial deposits. The excess pore-water pressure growth with increasing energy level makes the alluvium and the fractured-weathered rock less stiff and more dissipative, which reduces the PGA amplification. The PGA amplification factor for the AQM horizontal component (Figure 9d) and the AQG, AQA, AQV and AQM vertical components (Figures 9e, f, g and h) increases when the earthquake magnitude increases, which does not appear to be particularly influenced by the nonlinear behavior.

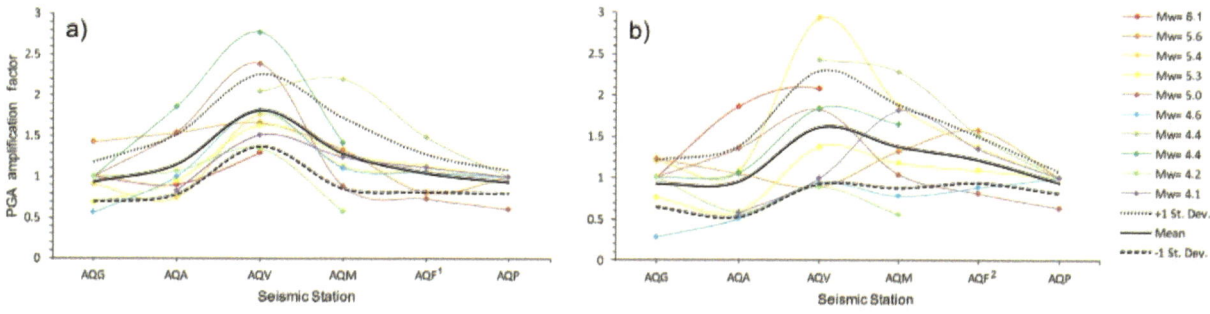

Fig. 8. Horizontal (a) and vertical (b) PGA amplification factors computed at seismic stations AQG, AQA, AQV, AQM, AQF and AQP using AQG and AQP as the reference sites for the 2009 events with magnitude M_w

4.1-6.1. M_W is the– moment magnitude and PGA is the– peak ground acceleration. 1 and 2 – AQF = (accepting the value) computed for the recorded M_W 5.6 event; the amplification factors are estimated.

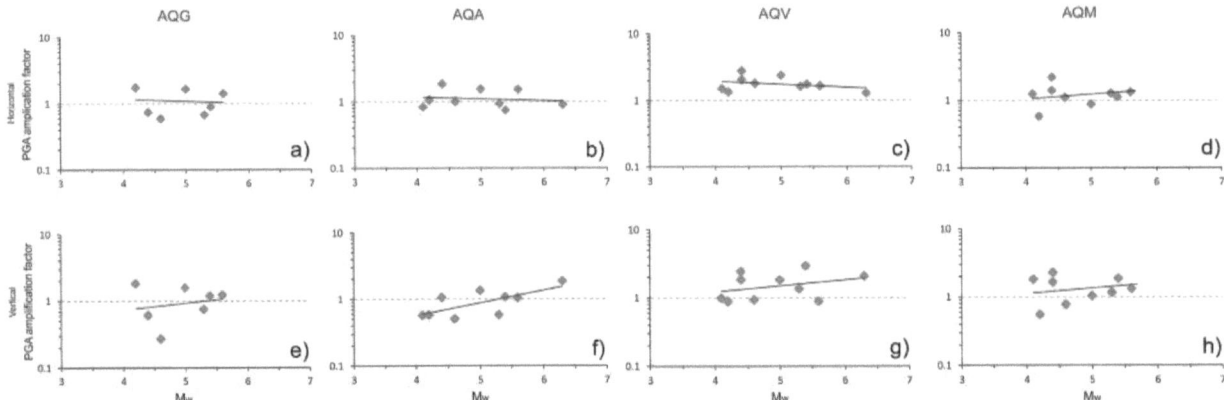

Fig. 9. Horizontal (a, b, c, d) and vertical (e, f, g, h) PGA amplification factors computed at seismic stations AQG, AQA, AQV and AQM as a function of the magnitude M_W. M_W is the– moment magnitude and PGA is the– peak ground acceleration.

Observed vs. Numerical Peak Ground Acceleration

To understand the peak values of acceleration that may occur at the seismic stations, during an earthquake such as the 2009 M_w 6.1 main shock, 102 nonlinear dynamic simulations are performed. We assume 10-second windows of the horizontal and vertical acceleration time-histories, which are recorded by AQG during the 2009 M_w 6.1 main shock and by AQG, AQM, AQP during the 2009 M_w 4.4-5.6 aftershocks, as the input motion records. The horizontal and vertical records are scaled to the peaks of 0.2 g and 0.1 g, respectively.

Table 4 presents a comparison between the observed horizontal maximum acceleration and the computed value for stations AQG, AQA, AQV and AQM. The values are notably consistent. The simulations reveal that during an event such as the 2009 M_w 6.1 main shock, the expected maximum acceleration is usually higher at the stations on soil and can vary at each station because the shape and the frequency content of the input seismic signal vary.

The 2D simulations have also allowed us to draw contour maps of the horizontal acceleration (Figure 10a) and the velocity (Figure 10b) and obtain the relative acceleration and velocity vectors, which display a clear seismic wave amplification at station AQV.

Table 4
Observed during the 2009 M_W 6.1 L'Aquila earthquake and computed (mean +/- 1 standard deviation) maximum acceleration at seismic stations AQG, AQA, AQV and AQM. NS denotes the– north-south component, and WE denotes the– west-east component.

Seismic Station				AQG	AQA	AQV	AQM
	Observed	NS		0.49	0.44	0.55	–
	(M_W 6.1 main shock)	WE		0.45	0.40	0.66	–
PGA (g)		- 1 St. Dev.		0.41	0.54	0.50	0.28
	Computed	Mean		0.49	0.59	0.59	0.38
		+ 1 St. Dev.		0.57	0.65	0.68	0.47

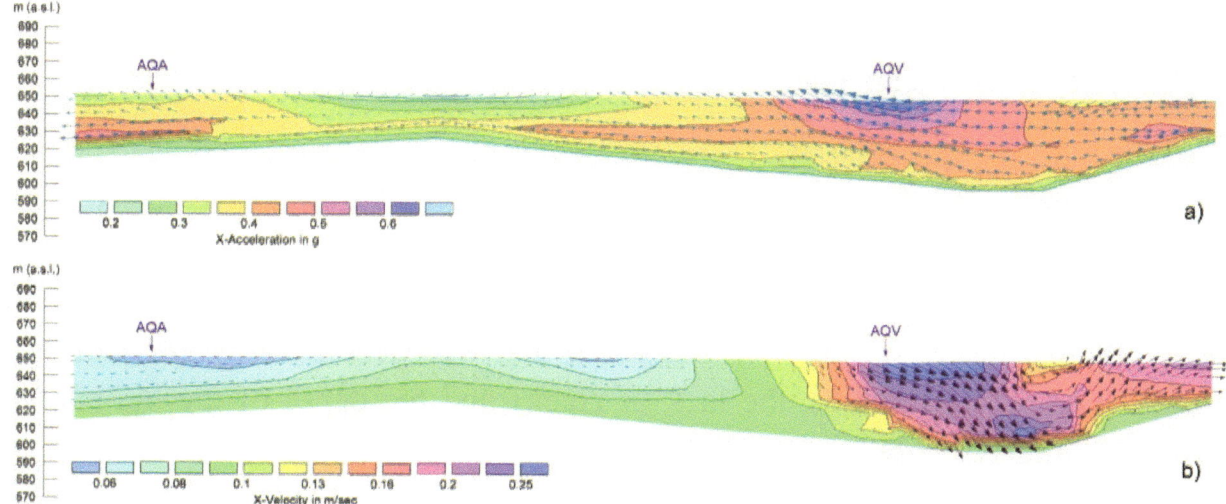

Fig. 10. 2D simulation of the 2009 M_W 6.1 L'Aquila earthquake using the model displayed in Figure 6b: a) X-acceleration contour map and the relative acceleration vectors (blue arrows) at 2.56 seconds of the nonlinear dynamic analysis; vector length magnification: 37 times; b) X-velocity contour map and the relative velocity vectors (black arrows) at 3.25 seconds of the nonlinear dynamic analysis; vector length magnification: 71 times. AQG,…, AQM are the– seismic stations.

Excess Pore-Water Pressure and Liquefaction

The 2D nonlinear dynamic analyses based on the MFS (*Martin-Finn-Seed*) pore-water pressure model [24] provided insights on the excess pore-water pressure that was generated under dynamic loading. In particular, the simulation of 2009 M_W 6.1 main shock revealed a maximum excess pore-water pressure of 400 kPa near the interface alluvium-carbonate bedrock and next to station AQV (Figure 11a). Based on the initial state of stress and the pore-water pressure changes, which were computed during the shaking, it was possible to evaluate the potential liquefaction zones (Figure 11b) in the units C, D1 and D2 (Figures 3, 4, 5 and 6). Moreover, the time-histories of excess pore-water pressure, which were computed at different depths under the seismic stations AQG, AQA and AQV (Figures 12a, b and c, respectively), exhibited a clear increase that reached the highest values after 2 to 3 seconds of shaking. The maximum excess pore-water pressure values, which range between 50 and 300 kPa, denote a reduction of the effective stress and shear resistance that is compatible with the observed fundamental frequency variation.

In addition, the graphs XY-shear stress vs. XY-shear strain (Figure 13), which are computed at ground level using the simulations of three events with magnitude 2.6, 4.6 and 6.3, corroborate the nonlinear seismic response because of the generation of excess pore-water pressure. In particular, the soil at stations AQA and AQV exhibits a strong nonlinear behavior during the 2009 M_w 6.1 L'Aquila earthquake simulation (Figures 13f and i). However, stations AQG (Figures 13a, b, and c) and AQM (Figures 13 l, m, and n) display a moderate nonlinearity at the highest magnitude.

In reality, during the 2009 M_w 6.1 main shock, liquefaction phenomena such as sand volcanos or blows occurred to the south-east of L'Aquila City [12,30], but they were not observed at the alluvial ground surface along the upper Aterno River Valley. However, (this occurrence cannot exclude) that liquefaction could really happen in the subsurface. The cover that was generated by geotechnical unit A (Figures 3, 4 and 5), the lack of cracks and the depth of the liquefiable granular soils possibly prevented the liquefaction effects from manifesting at ground level of the upper Aterno River Valley.

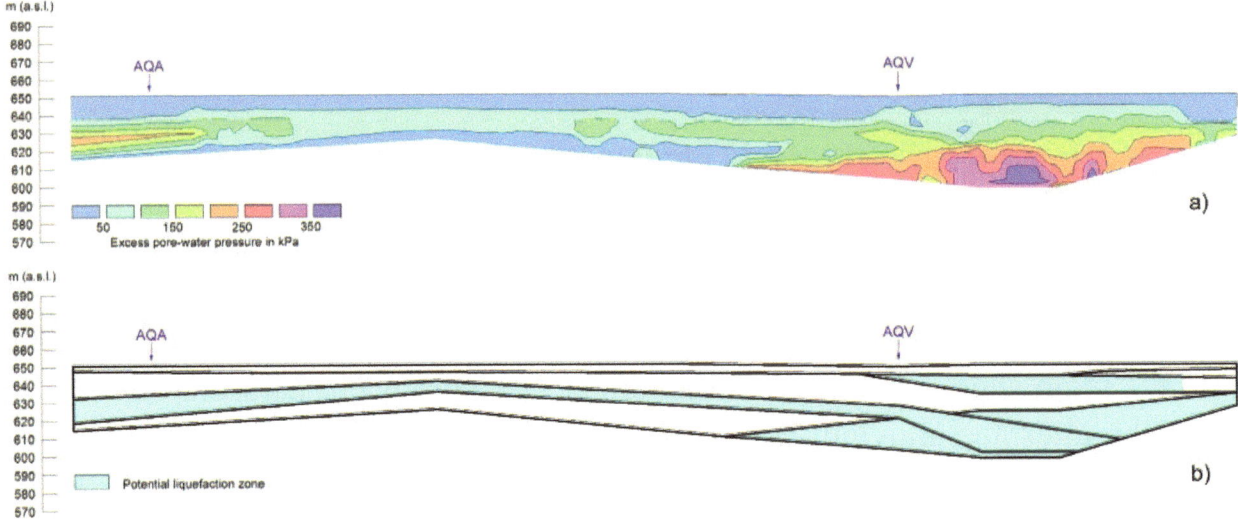

Fig. 11. 2D simulation of the 2009 M_w 6.1 L'Aquila earthquake using the model displayed in Figure 6b: a) excess pore-water pressure contour map at 3-4 seconds of the nonlinear dynamic analysis; b) potential liquefaction zone at 3-4 seconds of the nonlinear dynamic analysis. AQG,…, AQM are the– seismic stations.

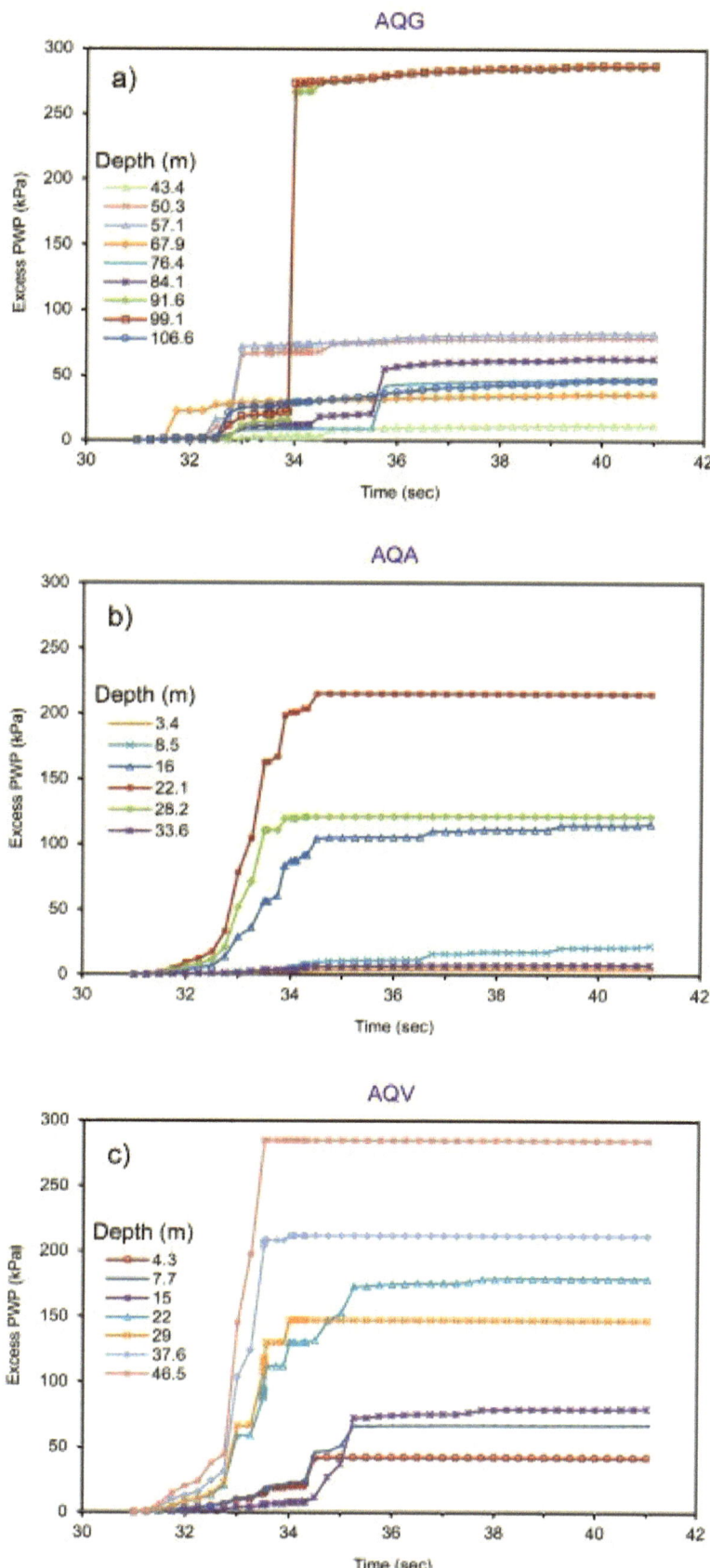

Fig. 12. 2D simulation of the 2009 M_W 6.1 L'Aquila earthquake using the models displayed in Figures 6a and b: time histories of the excess pore-water pressure (PWP) that were computed at seismic stations AQG (a), AQA (a) and AQV (b) for different depths.

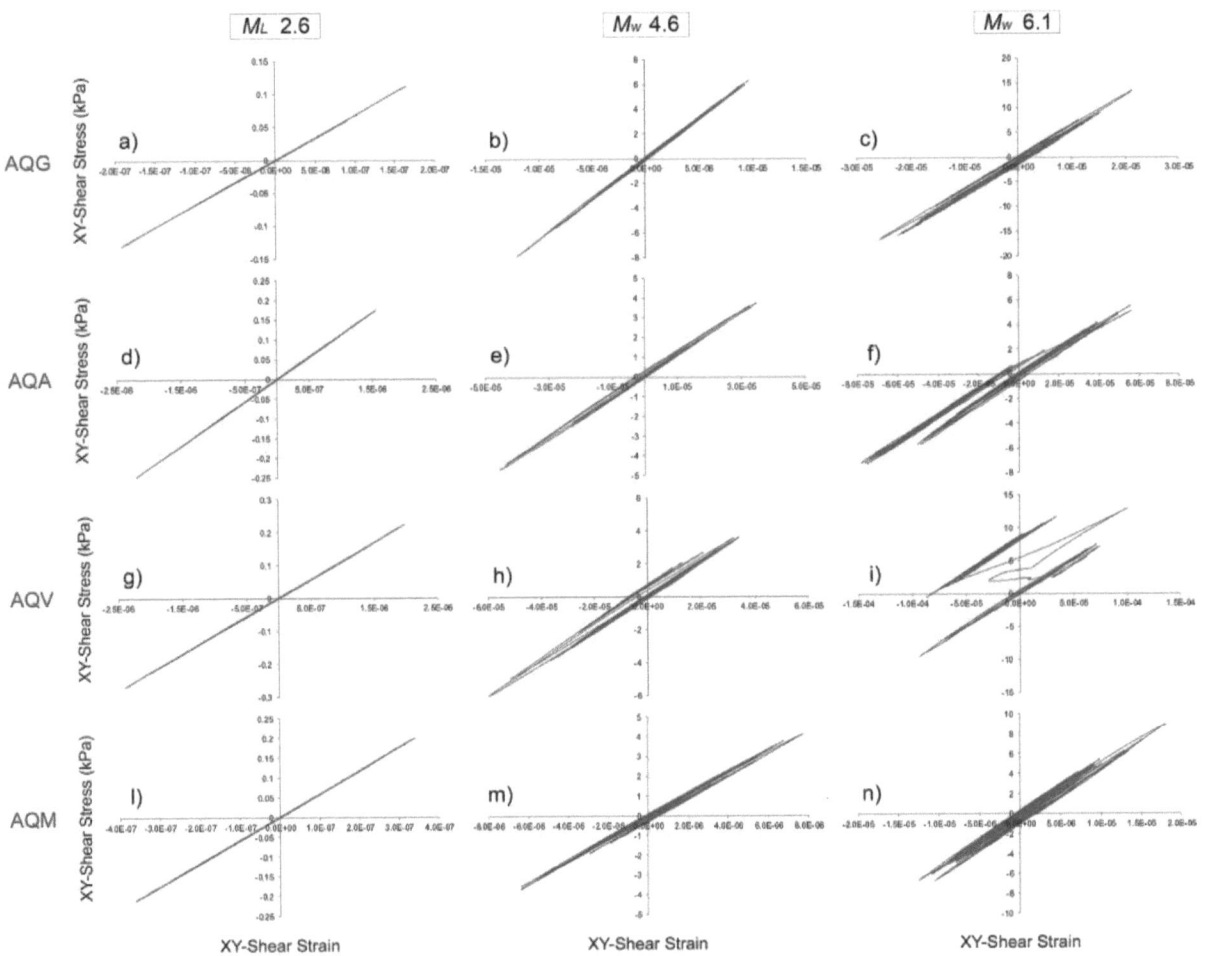

Fig. 13. XY-shear stress vs. XY-shear strain graphs, which were computed using the 2D nonlinear dynamic analysis for three events with magnitude 2.6, 4.6 and 6.1 at the ground surface at seismic stations AQG (a, b, c), AQA (d, e, f), AQV (g, h, i) and AQM (l, m, n). M_L is the– local magnitude and M_W is the– moment magnitude.

Observed vs. Numerical H/V and V/H Response Spectral Ratios

The H/V and V/H response spectral ratios from the ground motion records analysis (Figures 14, 15, 16 and 17) and the 2D nonlinear dynamic modeling (nonlinear analyses in Figures 14, 15, 16 and 17), provided a better understanding of the soil and rock dynamic behavior at seismic stations AQG, AQA, AQV and AQM.

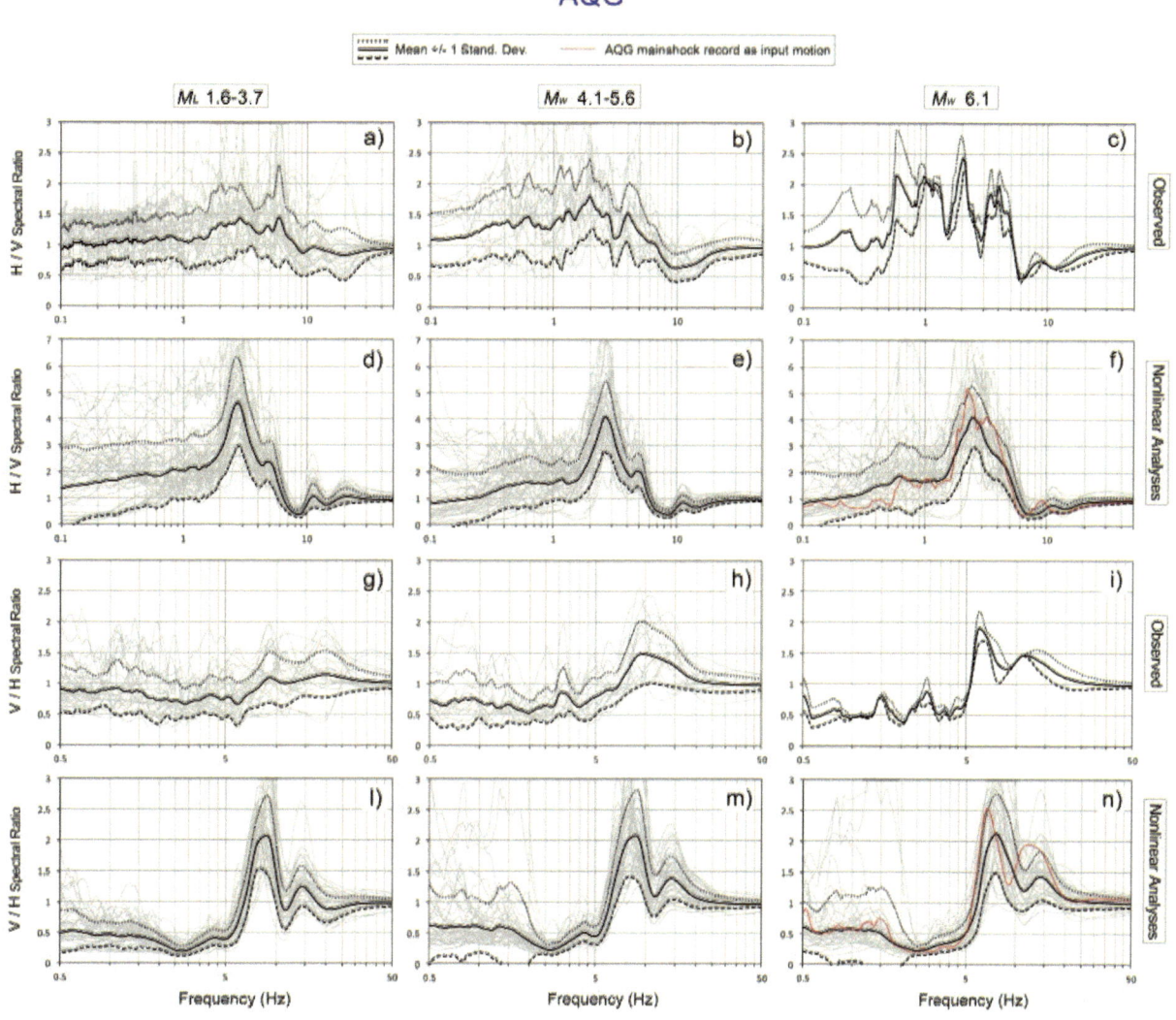

Fig. 14. Observed (a, b, c, g, h, i) and computed (d, e, f, l, m, n) H/V and V/H spectral ratios using the 2D nonlinear dynamic model for three earthquake magnitude levels at the ground surface at seismic station AQG. M_L is the– local magnitude and M_W is the– moment magnitude.

At AQG, for the lower earthquake magnitude level (M_L 1.6-3.7), the observed H/V spectral ratio exhibits two prevalent peaks at (Figure 14a): (the first peak is at 5-6 Hz, which can be related to the superposition of the highly fractured and weathered rock unit F1(2) on the fractured rock unit F1 at a depth of 25 meters (Figure 5a); the second peak is at 2-3 Hz, which can be related to a deeper seismic impedance contrast at the interface between units F1 and G in Figure 3) (13,18,21,23,27,31,39). For higher earthquake magnitudes (M_w 4.1-6.1), the observed H/V spectral ratio displays two evident peaks at 2 Hz and 4 Hz (Figures 14b and c), which denote a resonance frequency reduction of up to 30 % and are probably related to the nonlinear response of the highly fractured and weathered rock (F1 and F1(2) in Figures 4 and 5a). Two peaks are also revealed at 0.6 Hz and 1 Hz (Figures 14b and c). The nonlinear response most likely also increased the level of maximum amplification from 1.5 to 2.5 when the earthquake magnitude increased (Figures 14a, b and c).

The observed frequencies of maximum amplification and the variation are also displayed by the numerical H/V spectral ratio (Figures 14d, e and f) using 2D nonlinear modeling. Thus, the assumed subsurface model (Figures 3 and 6a) and the nonlinear seismic effects at station AQG are reliable.

Finally, the observed V/H spectral ratio (Figures 14g, h and i) and the numerical values (Figures 14 l, m and n) exhibit maximum amplification at frequencies of 10 Hz and 6 Hz, which are almost double the values (of the H/V ratio) and the maximum amplification level increases when the earthquake magnitude increases.

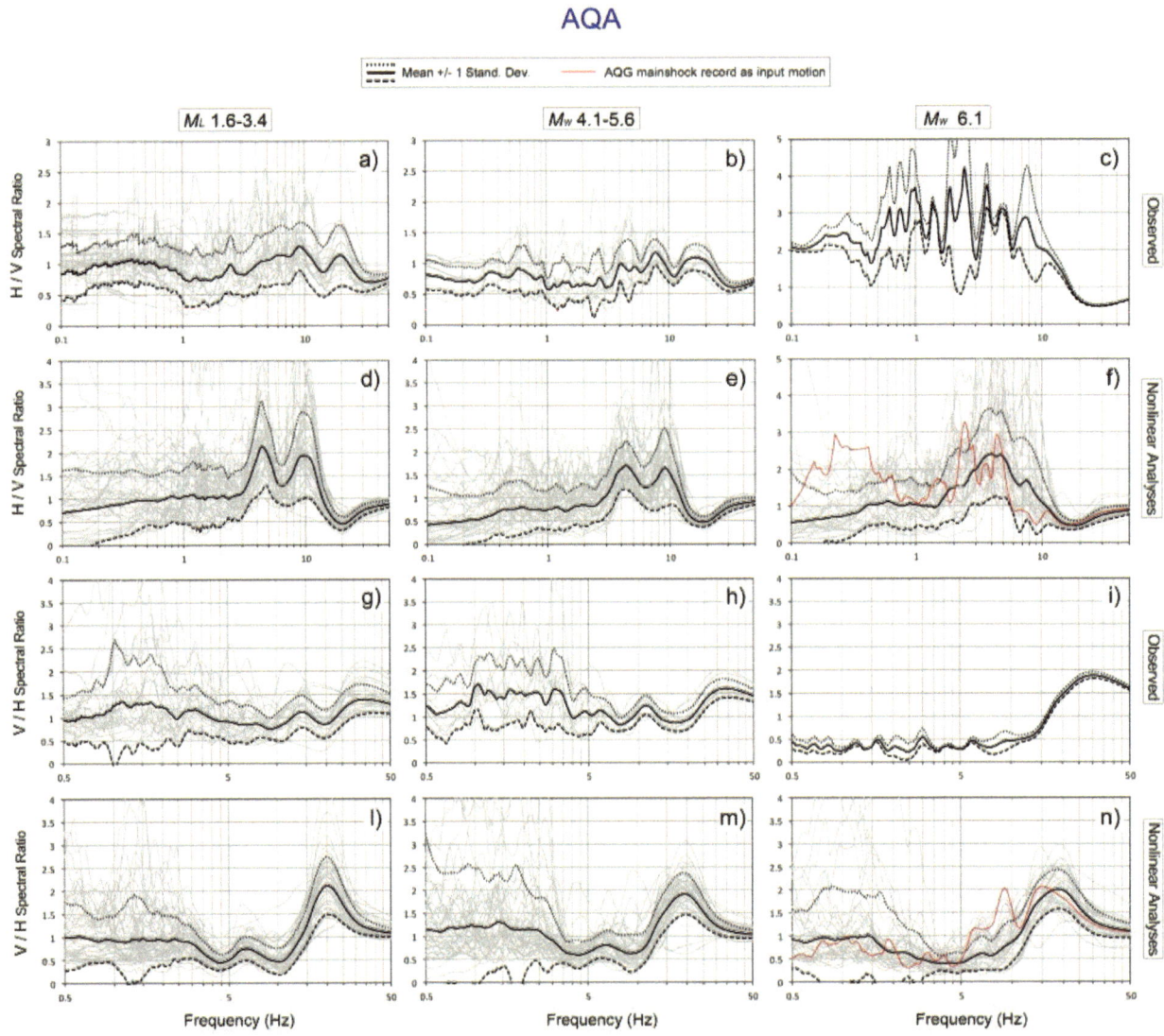

Fig. 15. Observed (a, b, c, g, h, i) and computed (d, e, f, l, m, n) H/V and V/H spectral ratios using the 2D nonlinear dynamic model for three earthquake magnitude levels at the ground surface at seismic station AQA. M_L is the– local magnitude and M_W is the– moment magnitude.

At AQA, for the lower earthquake magnitudes (M_L 1.6-3.4), the observed H/V spectral ratio exhibits a maximum amplification at 20 Hz, 9 Hz and 4-6 Hz. The 20 Hz amplification (Figure 15a) disappears during the 2009 M_w 6.1 main shock. The 9 Hz amplification can be related to the superposition of alluvial unit A on unit B1 (Figures 5b, c and d). The 4-6 Hz amplification is related to the seismic impedance contrast, which occurs when alluvial unit B2 contacts rock unit F2 (Figures 3, 4 and 5d). (The alluvial soil has a remarkable nonlinear response, which is evidenced by the simultaneous reduction of the resonance frequency from 4-6 Hz to 2.5 Hz and the increase in the maximum amplification level to 4) during the 2009 M_w 6.1 main shock (Figure 15c). In addition, (a nonlinear response can occur by reducing the 9 Hz resonance frequency to 4-5 Hz) (Figure 15c).

The maximum amplification at 4-5 Hz and 9 Hz, which is observed at the lower earthquake magnitudes, is also exhibited in the numerical H/V spectral ratio from 2D nonlinear modeling. For example, the nonlinear response of the alluvial soil is simultaneously evidenced by the reduction in the 4-5 Hz resonance frequency to 2.5 Hz and the increase in the maximum amplification level to 3.5 (Figures 15d, e and f). The maximum amplification at 4-5 Hz, which is observed in Figure 15c, is also displayed by the 2009 M_w 6.1 simulation (Figure 15f).

The observed V/H spectral ratio displays a distinct maximum amplification at 30 Hz for all earthquake magnitude levels (Figures 15g, h and i) and a maximum amplification level ranging between 1.5 and 2. This amplification is related to the P-wave impedance contrast between alluvial units A and B1 (Figures 5b, c and d).

Amplification is also exhibited at 10 Hz, 15 Hz and in the range of 1-3 Hz (Figures 15g and h), but it disappears during the 2009 M_w 6.1 event (Figure 15i).

The numerical V/H spectral ratio also shows an evident high-frequency amplification at 20 Hz with a mean maximum amplification level of 2 for all earthquake magnitude levels (Figures 15 l, m and n).

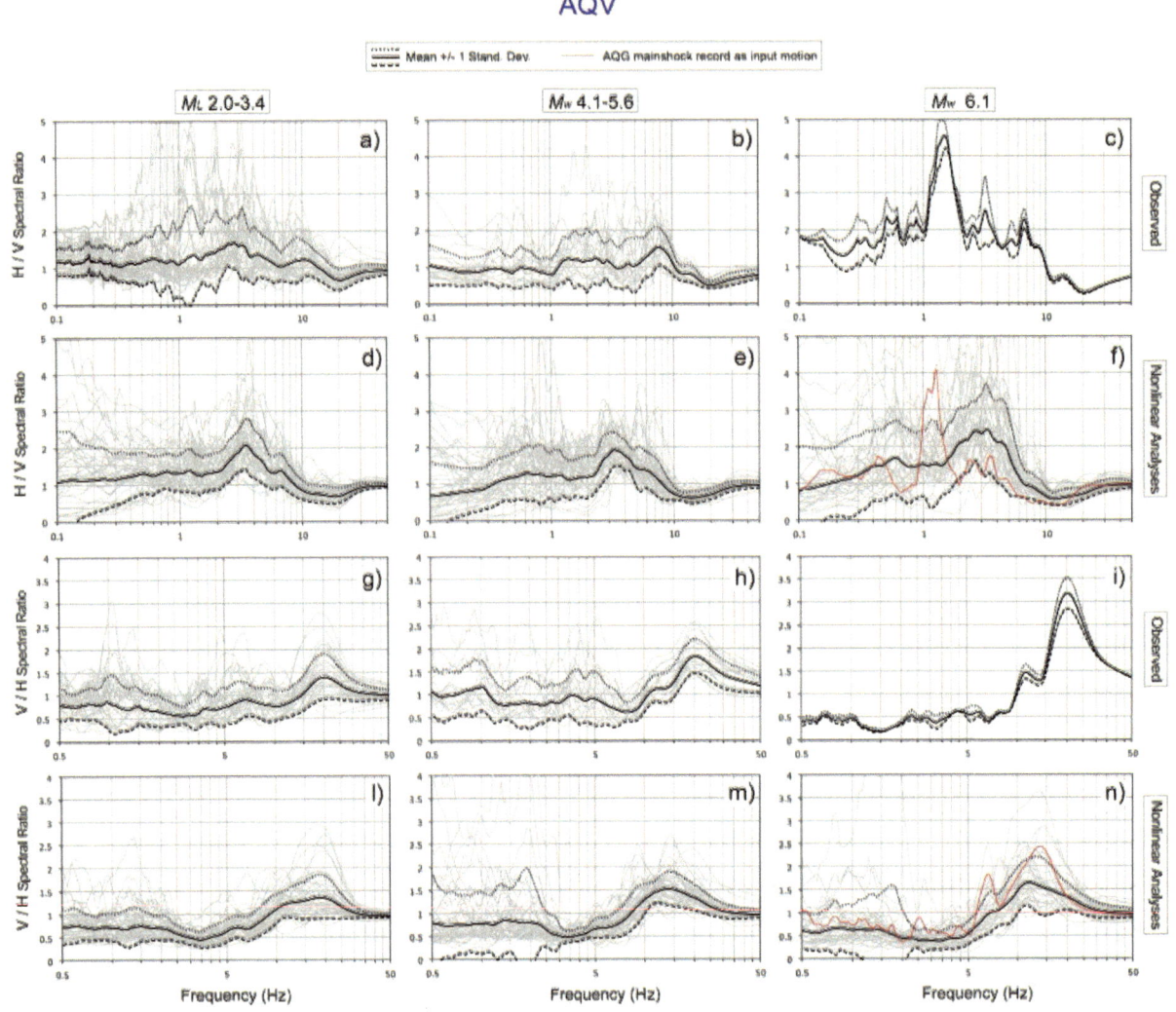

Fig. 16. Observed (a, b, c, g, h, i) and computed (d, e, f, l, m, n) H/V and V/H spectral ratios using the 2D nonlinear dynamic model for three earthquake magnitude levels at the ground surface at seismic station AQV. M_L is the– local magnitude and M_w is the– moment magnitude.

At AQV, for the lower earthquake magnitude level (M_L 2.0-3.4), the observed H/V spectral ratio exhibits maximum amplification at 10 Hz (Figure 16a), which can be related to the seismic impedance contrast between alluvial units A and B1 (Figures 5b, c and d), and 3 Hz, which can be related to the superposition of alluvial unit D2 on rock unit F2 (Figure 5e). In addition, in this case, it is interesting to note the nonlinear response of the alluvial soil, which is simultaneously evidenced by the decrease in the 3 Hz resonance frequency to 1.5 Hz and the increase in the mean maximum amplification level from 1.5 to 4.5, when the earthquake magnitude increases (Figures 16a, b and c).

The maximum amplification at 3 Hz, which is observed at the lower earthquake magnitudes, is also clearly exhibited by the numerical H/V spectral ratio from 2D nonlinear modeling, such as the nonlinear response of the alluvial soil. This response is simultaneously evidenced by the reduction in the 3 Hz resonance frequency to 1.5 Hz and the increase in the maximum amplification level to 4 when the earthquake magnitude increases (Figures 16d, e and f). In particular, for both seismic stations AQV and AQA, it is remarkable that only the 2D simulation could generate similar spectral amplification functions (red functions in Figures 15f and 16f) to the observed ones (Figures 15c and 16c) in terms of the frequency and level of maximum amplification. (For the input motion records, the 2D simulation used the signals that were collected by the AQG station during the 2009 M_w 6.1 main shock).This result can be explain by the frequency content and the higher duration and amplitude of the most energetic seismic phases, (which characterize the 2009 M_w 6.1 signals recorded by AQG and further increase the plastic volumetric strain and the excess pore-water pressure) (Martin et al., 1975).

The observed V/H spectral ratio displays a constant maximum amplification at 20 Hz for all earthquake magnitude levels (Figures 16g, h and i), which is related to the P-wave impedance contrast between alluvial units A and B1 (Figures 5b and c), whereas the mean level of maximum amplification gradually increases from 1.5 to 3 when the earthquake magnitude increases.

The numerical V/H spectral ratio also shows a high-frequency amplification at 15-20 Hz, whose level increases from 1.5 to 2.5 when the earthquake magnitude reaches M_w 6.1 (Figures 16 l, m and n).

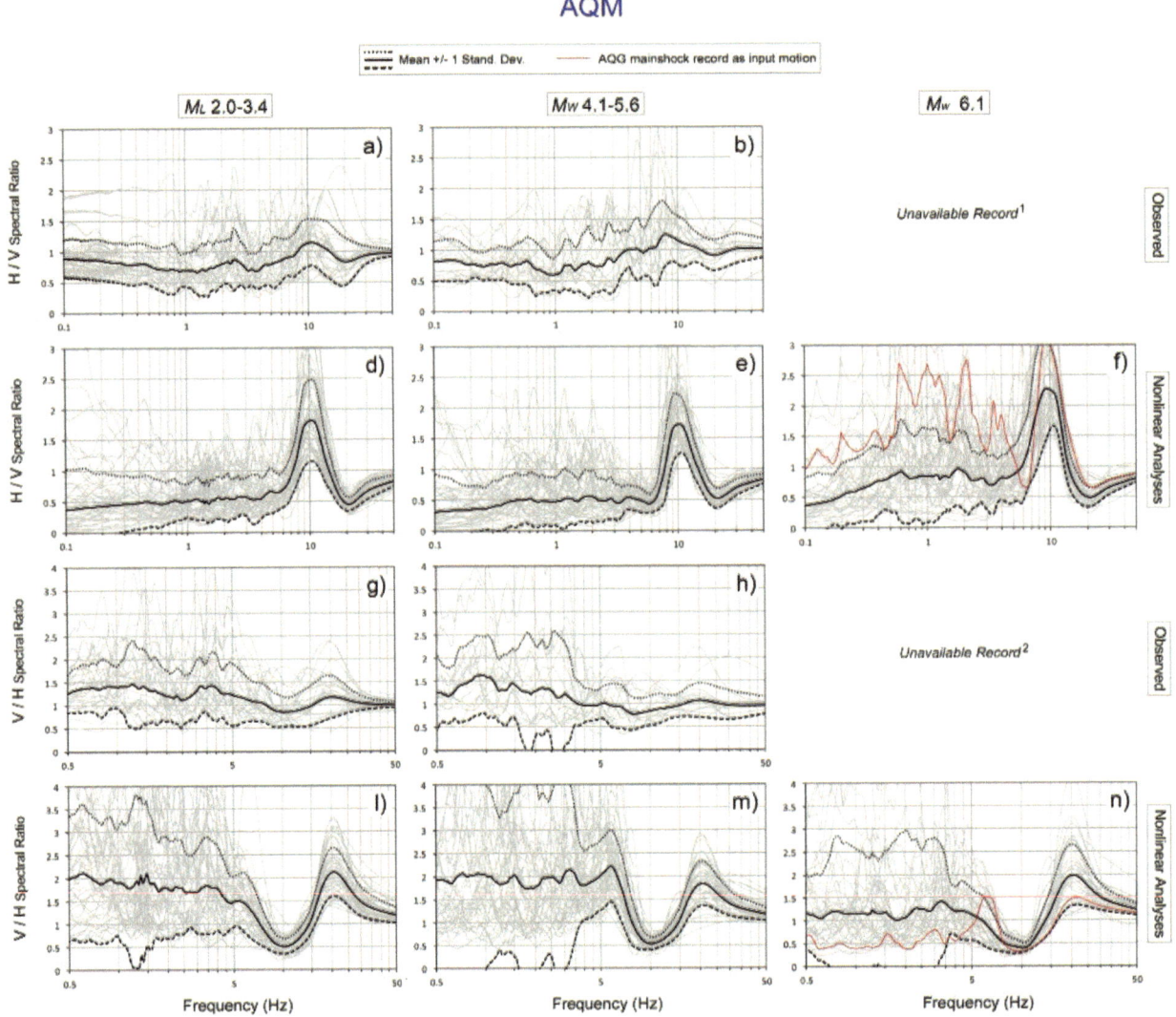

Fig. 17. Observed (a, b, g, h) and computed (d, e, f, l, m, n) H/V and V/H spectral ratios using the 2D nonlinear dynamic model for three earthquake magnitude levels at the ground surface at seismic station AQM. M_L is the– local magnitude and M_W is the– moment magnitude.

1 and 2 – *Unavailable Record* = AQM recorded the M_W 6.1 main shock, but the data were not formally released because the recorded motion could be disturbed by earthquake-induced damage to the seismometer protective box (Akinci et al., 2010).

At AQM, the observed and computed H/V spectral ratios exhibit a significant response at 10 Hz for all earthquake magnitude levels, and the mean level of maximum amplification that increases from approximately 1.5 to 3 when the earthquake magnitude increases (Figures 17a, b, d, e, and f). The maximum amplification at 10 Hz can be related to the seismic impedance contrast between the carbonate bedrock (Ma in Figure 2 and F3 in Figures 3 and 4) and the shallow strata of fractured/weathered limestone (5 in Figure 3) and detrital alluvial deposits (Cd in Figure 2 and E in Figure 3).

Finally, both the observed and the numerical V/H spectral ratios exhibit a clear response at 20 Hz, which is double the horizontal resonance frequency, and in the frequency band 0.5-5 Hz, and the mean level of maximum amplification does not notably vary when the earthquake magnitude increases (Figures 17g, h, l, m, and n).

Conclusions

The seismic site response at four near-fault stations of the Aterno River Valley strong-motion array, which is placed on both soil and rock outcrops of Western L'Aquila Basin (central Italy), is studied by analyzing 312 seismic signals and 366 2D FEM numerical simulations. The signals were mostly recorded during the 2009 L'Aquila seismic sequence and the numerical simulations were performed on three models, (which were drawn from a geologic cross-section that passed through the array).

The strong- and weak-motion record analyses show the following:
- The seismic motion amplification is usually higher at station AQV, which is located near the valley center in the thickest part of the alluvial deposit.

- The observed frequency of maximum amplification for lower earthquake magnitude levels can be related to the depth of impedance contrast between the geotechnical units, (as revealed from the 2D subsurface model).

- The vertical component of the seismic motion exhibits a high-frequency amplification at 20-30 Hz at stations AQA, AQV and AQM for all earthquake magnitude levels.

- The nonlinear site response is observed for the horizontal component at station AQG on highly fractured-weathered rock and AQA and AQV on alluvial soil, as revealed from i) the decrease in the PGA amplification factor when the earthquake magnitude increases; and ii) the decrease in the fundamental frequency (up to 30% at AQG and 50% at AQA and AQV) and the increase in the maximum amplification level when the earthquake magnitude increases.

The results of the 2D numerical simulations based on the finite element method are consistent with the experimental results and show the following:

- During an event such as the 2009 M_w 6.1 main shock, the expected peak acceleration is usually higher at the stations on soil and can vary at each station because the shape and the frequency content of the input seismic signal vary.

- The observed nonlinear seismic response at stations AQG, AQA and AQV can be related to the generation of excess pore-water pressure, which ranges between 50 and 300 kPa during the 2009 M_w 6.1 main shock.

The 2D dynamic modeling based on the Martin-Finn-Seed's pore-water pressure model was performed in this study on a complex two-dimensional irregular configuration. The model considered several important parameters such as the rock and soil elastic and dynamic properties, the intensity and the depth of the seismic impedance ratio, the water table depth, the dip angle of strata, the magnitude, the shape and the frequency content of the input seismic signal.

Previous studies attempted to simulate the seismic site response at station AQV, but they could not completely explain the recorded motions. Using the 2D finite element nonlinear dynamic model, this paper can correctly estimate the observed frequency variations and levels of maximum amplification at stations AQG, AQA and AQV and the observed seismic site response at station AQM.

References

[1] Ameri G, Massa M, Bindi D, D'Alema E, Gorini A, Luzi L, et al. The 6th April 2009 Mw 6.3 L'Aquila (central Italy) earthquake: strong-motion observations. Seismological Research Letters 2009;80(6):951–66.

[2] Amoroso S, Del Monaco F, Di Eusebio F, Monaco P, Taddei B, Tallini M, Totani F, Totani G. Campagna di indagini geologiche, geotecniche e geofisiche per lo studio della risposta sismica locale della città dell'Aquila: la stratigrafia dei sondaggi (giugno–agosto 2010). Report CERFIS n. 1 2010; http//www.cerfis.it/en/download/cat_view/67-pubblicazioni-cerfis/68-reports.

[3] Akinci A, Malagnini L, Sabetta F. Characteristics of the strong ground motions from the 6 April 2009 L'Aquila earthquake, Italy. Soil Dyn Earthq Eng 2010;30:320–335.

[4] Bongiovanni G, Gorini A, Gorelli V, Marcucci S, Marsan P, Milana G. Primi risultati della rete accelerometrica locale dell'Aquila e sistema di monitoraggio del Sacro Convento di Assisi. Rapporto tecnico 1995;SSN/RT/95/6: 5–19.

[5] Bindi D, Pacor F, Luzi L, Massa M, Ameri G. The Mw 6.3, 2009 L'Aquila earthquake:source, path and site effects from spectral analysis of strong motion data. Geophysical Journal International 2009;179:1573-1579.

[6] Bonilla LF, Archuleta RJ, Lavallée D. Histeretic and dilatant behavior of cohesionless soils and their effects on non-linear site response: field data observation and modelling. Bull Seism Soc Am 2005;95:2373-2395.

[7] Byrne PM. A cyclic shear-volume coupling and pore pressure model for sand. Proceedings, 2nd international conference on Recent Advances in Geotechnical Earthquake and Soil Dynamics, St. Louis, Missouri, March 1991;1:47-56.

[8] Chioccarelli E, Iervolino I. Near Source seismic demand and pulse-like record: A discussion for L'Aquila earthquake. Earthq Eng Struct Dyn 2010;39:1039-1062.

[9] Chopra AK. Dynamics of Structures: Theory and Applications to Earthquake Engineering. Englewood Cliffs, NJ: Prentice Hall; 1995.

[10] De Luca G, Marcucci S, Milana G, Sanò T. Evidence of low-frequency amplification in the city of L'Aquila, central Italy, through a multidisciplinary approach including strong and weak motion data, ambient noise, and numerical modeling. Bull Seism Soc Am 2005;95:1469–1481.

[11] De Martin F, Kawase H, Modaressi-Farahmand Razavi A. Nonlinear Soil Response of a Borehole Station Based on One-Dimensional Inversion during the 2005 Fukuoka Prefecture Western Offshore Earthquake. Bull Seism Soc Am 2010;100:151–171.

[12] De Martini PM, Cinti FR, Cucci L, Smedile A, Pinzi S, Brunori CA, Molisso F. Sand volcanoes induced by the April 6th 2009 Mw 6.3 L'Aquila earthquake: a case study from the Fossa area. Ital J Geosc 2012;131:410-422.

[13] Del Monaco F, Tallini M, De Rose C, Durante F. HVNSR survey in historical downtown L'Aquila (central Italy): Site resonance properties vs. subsoil model. Eng Geol 2013;158:34–47.

[14] Ditommaso R, Mucciarelli M, Ponzo FC. S-Transform based filter applied to the analysis of non-linear dynamic behaviour of soil and buildings. 14th European Conference on Earthquake Engineering. Proceedings Volume. Ohrid, Republic of Macedonia 2010; August 30 – September 3.

[15] Di Fiore V, Cavuoto G, Del Monaco F, Mancini M, Caielli G, Cavinato GP, De Franco R, Pelosi N, Rapolla A, Tallini M. Seismic surveys integrated with geological data for in-depth investigation of Mt. Pettino active fault area (Western L'Aquila Basin). Ital Journ Geosc 2012;131.

[16] Di Capua G, Lanzo G, Luzi L, Pacor F, Paolucci R, Peppoloni S, Scasserra G, Puglia R. Caratteristiche geologiche e classificazione di sito delle stazioni accelerometriche della RAN ubicate a L'Aquila. Project S4 (2007-2009): http://esse4.mi.ingv.it.

[17] Di Capua G, Lanzo G, Pessina V, Peppoloni S, Scasserra G. The recording stations of the Italian strong motion network: Geological information and site classification. Bull Earthq Engin 2011;9:1779-1796.

[18] Field EH, Jacob KH. A comparison and test of various site response estimation techniques, including three that are not reference site dependent. Bull Seism Soc Am 1995;85:1127–1143.

[19] Gaudiosi I, Del Monaco F, Milana G, Tallini M. Site effects in the Aterno River Valley (L'Aquila, Italy): comparison between empirical and 2D numerical modelling starting from April 6th 2009 Mw 6.3 earthquake. Bull Earthq Engin 2014;12:697-716.

[20] Gorini A, Nicoletti M, Marsan P, Bianconi R, De Nardis R, Filippi L, Marcucci S, Palma F, Zambonelli E. The Italian strong motion network. Bull Earthq Engin 2010;8:1075-1090.

[21] Huang HC, Teng TL. An evaluation on H/V ratio vs. spectral ratio for site-response estimation using the 1994 Northridge earthquake sequences. Pur Appl Geoph 1999;156:631–649.

[22] Kham M, Semblat JF, Bouden-Romdhane N. Amplification of seismic ground motion in the Tunis basin: Numerical BEM simulations vs experimental evidences. Eng Geol 2013;154:80–86.

[23] Konno K, Ohmachi T. Ground-motion characteristics estimated from spectral ratio between horizontal and vertical components of microtremors. Bull Seism Soc Am 1998;88:228–241.

[24] Kramer SL. Geotechnical earthquake engineering. Upper Saddle River, NJ: Prentice Hall; 1996.

[25] Lanzo G, Tallini M, Milana G, Di Capua G, Del Monaco F, Pagliaroli A, Peppoloni S. The Aterno Valley strong-motion array: seismic characterization and determination of subsoil model. Bull Earthq Eng 2011;9:1855–1875.

[26] Lanzo G, Pagliaroli A. Seismic site effects at near-fault strong-motion stations along the Aterno River Valley during the Mw = 6.3 2009 L'Aquila earthquake. Soil Dyn Earthq Eng 2012;40:1–14.

[27] Lermo J, FJ Chavez Garcia. Site effects evaluation using spectral ratios with only one station. Bull Seism Soc Am 1993;83:1574–1594.

[24] Martin GR, Finn WDL, Seed HB. Fundamentals of liquefaction under cyclic loading. J Geotech Eng Div ASCE, GT5 1975:423–38.

[28] Massa M, Pacor F, Luzi L, Bindi D, Milana G, Sabetta F, Gorini A, Marcucci S. The ITalian ACcelerometric Archive (ITACA): processing of strong-motion data. Bull Earthq Eng 2010;8:1175-1187.

[29] Mayeda K, Malagnini L, Walter WR. A new spectral ratio method using narrow band coda envelopes: evidence for non-selfsimilarity in the Hector Mine se- quence. Geophy Res Lett 2007;34:L11303.

[30] Monaco P, Santucci de Magistris F, Grasso S, Marchetti S, Maugeri M, Totani G. Analysis of the liquefaction phenomena in the village of Vittorito (L'Aquila). Bull Earthq Eng 2010;9:231-261.

[31] Nakamura Y. A method for dynamic characteristics estimation of subsurface using microtremor on the ground surface. Q Rept Railw Tech Res Inst 1989;30:1.

[32] Newmark NM. A method of computation for structural dynamics. ASCE J Eng Mech Div 1959;85 No EM3.

[33] Nunziata C, Costanzo MR. Ground motion modeling for site effects at L'Aquila and middle Aterno river valley (central Italy) for the Mw 6.3, 2009 earth- quake. Soil Dyn Earthq Eng 2014;61–62:107–23.

[34] Puglia R, Ditommaso R, Pacor F, Mucciarelli M, Luzi L, Bianca M. Frequency variation in site response as observed from strong motion data of the L'Aquila (2009) seismic sequence. Bull Earthq Eng 2011;9:869–92.

[35] QUAKE/W. Geo-slope international Ltd. engineering book. 3rd ed. 2008 Version 2007.

[36] Ragozzino E. Nonlinear seismic response in the western L'Aquila basin(Italy): numerical FEM simulations vs. ground motion records. Eng Geol 2014;174:46– 60.

[37] Ragozzino E. Seismic response of deep Quaternary sediments in historical center of L'Aquila City (central Italy). Soil Dyn Earthqu Eng 2016;87:29–43.

[38] Sabetta F, Rovelli A, Celebi M, Rinaldis D. Sequenza sismica dell'Abruzzo: analisi delle registrazioni accelerometriche. En Amb Innov 2009;3:12–27.

[39] Sawada Y, Taga M, Watanabe M, Nakamoto T, et al. Applicability of micro- tremor H/V method for KIK-NET strong motion observation sites and Nobi plain. In: Proceedings of the13th World Conference on Earthquake Engineering 2004, Vancouver (B.C., Canada); 1–6 August 2004. Paperno. 855.

[40] Sawazaki K, Sato H, Nakahara H, Nishimura T. Temporal change in site re- sponse caused by earthquake strong motion as revealed from coda spectral ratio measurement. Geophys Res Lett 2006;33:L21303.

[41] Tanimoto T, Ji C, Archuleta R. Inversion and prediction of ground motion of the 2009 L'Aquila Italy Mw 6.3 Earthquake. USGS; 2011. Award Number G10AP00010 2010/1/1-2011/12/3.

[42] Theodulidis N, Bard PY. Horizontal to vertical spectral ratio and geological conditions: an analysis of strong motion data from Greece and Taiwan (SMART-I). Soil Dyn Earthq Eng 1995;14:177–97.

[43] Theodulidis N, Bard PY, Archuleta R, Bouchon M. Horizontal-to-vertical spectral ratio and geological conditions: the case of Garner Valley downhole array in Southern California. Bull Seism Soc Am 1996;86:306-319.

[44] Totani G, Monaco P, Marchetti S, Marchetti D. Vs measurements by seismic dilatometer (SDMT) in non-penetrable soils. In: Proceedings of 17th ICSMGE 2009; vol 2. Alexandria, Egypt: 977–980.

[45] Vezzani L, Ghisetti F. Carta geologica dell'Abruzzo (Scala1:100.000). S.EL.CA., Firenze; 1998.

[46] Working Group ITACA, 2010. Data Base of the Italian strong motion records: http://itaca.mi.ingv.it.

[47] Working Group MS–AQ. Microzonazione sismica per la ricostruzione dell'area aquilana. L'Aquila: Regione Abruzzo – Dipartimento della Protezione Civile 2010; vol 3 and Cd-rom.

[48] Wu C, Peng Z, Ben-Zion Y. Non-linearity and temporal changes of fault zone site response associated with strong ground motion. Geophys J Int 2009;176:265–78.

[49] Wu C, Peng Z, Assimaki D. Temporal Changes in Site Response Associated with the Strong Ground Motion of the 2004 Mw 6.6 Mid-Niigata Earthquake Sequences in Japan. Bull Seism Soc Am 2009;99:3487-3495.

[50] Wu G. Volume change and residual pore water pressure of saturated granular soils to blast loads. A research report submitted to Natural Sciences and En- gineering Research Council of Canada; 1996.

[51] Zambonelli E, De Nardis R, Filippi L, Nicoletti M, Dolce M. Performance of the Italian strong motion network during the 2009, L'Aquila seismic sequence (central Italy). Bull Earthq Eng 2010;9:39-65.

Case Study #2: Historical L'Aquila City center

Geological, Geotechnical and Geophysical Data

The medieval City of L'Aquila sits upon a hillside in the middle of a narrow and NW-SE trending intramontane valley, through which the Aterno River flows (Fig. 1).

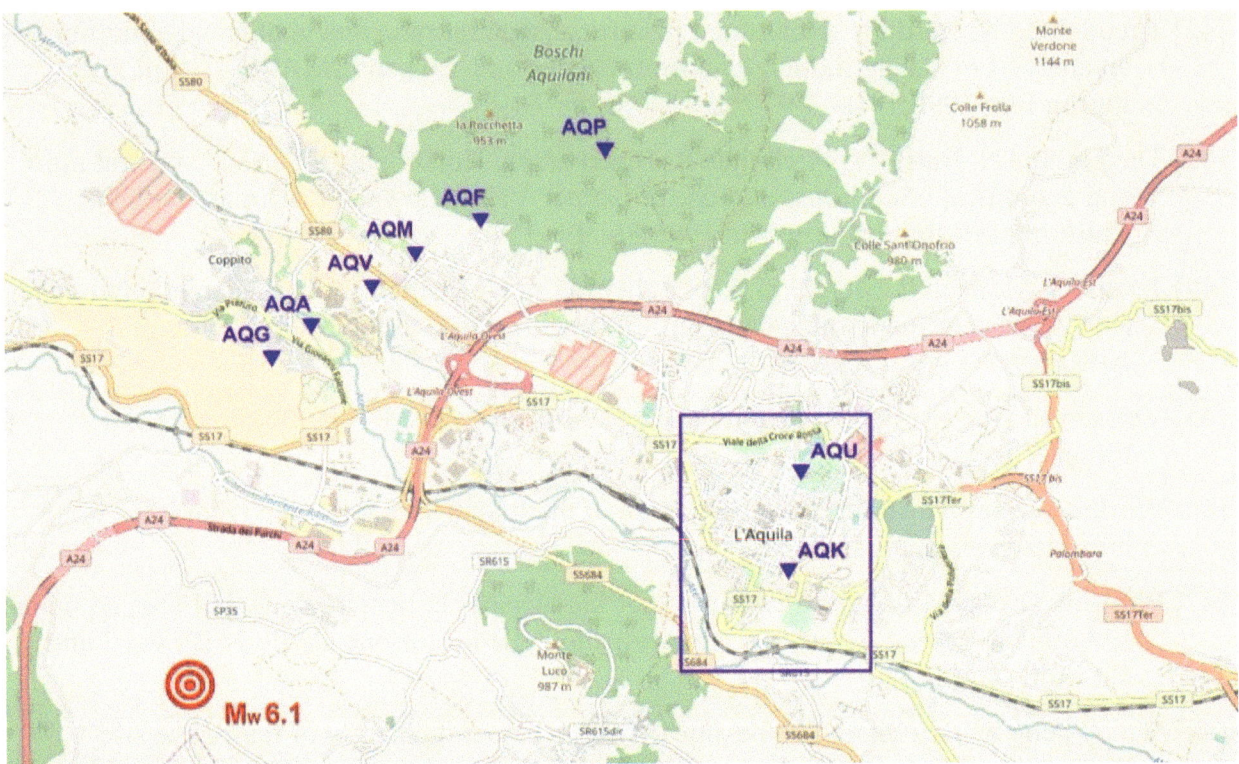

Fig. 1. Study area map showing the position of seismic stations AQK and AQU as well as the Aterno River Valley strong-motion array and the epicenter of 2009 M_w 6.1 main shock.
Map © OpenStreetMap contributors, http://www.openstreetmap.org/copyright/en

The valley was generated by the extensional tectonics, which characterized the central part of the Apennines since the Pliocene and produced mainly NW-SE and NE-SW trending normal faults.

The geologic setting consists of Quaternary fluvial, lacustrine [18] and slope sediments that fill the Aterno River Valley and overlay a variously dislocated Meso-Cenozoic carbonate bedrock [13,38,41,47,48].

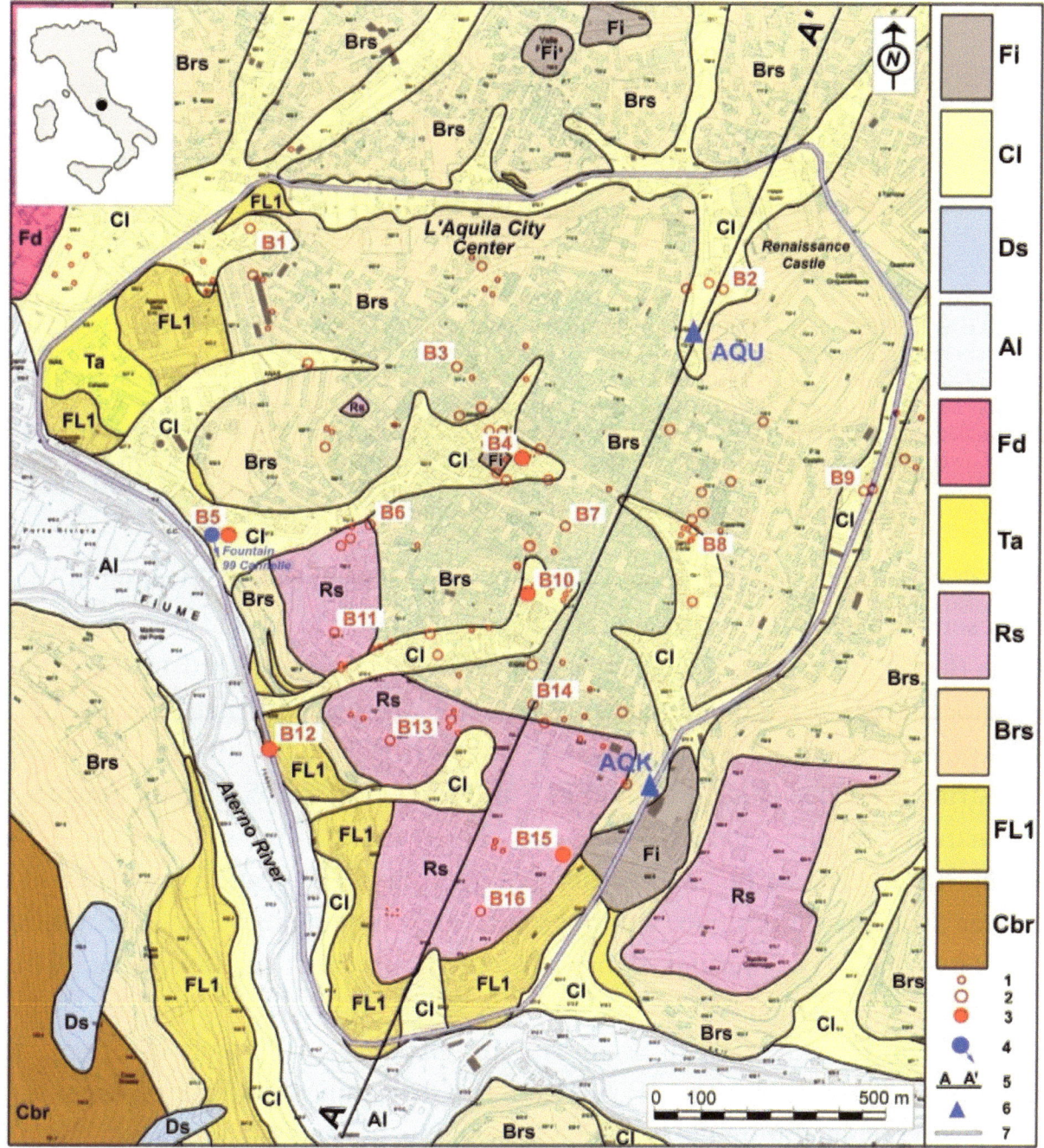

Fig. 2. Geologic map of medieval L'Aquila City. Quaternary units: Fi – filling anthropogenic material (Holocene); Al – Aterno river alluvium: coarse- and fine-grained deposits (Holocene); Cl – colluvium: sandy silty gravel, gravelly sandy silt and silty sand (Holocene); Ds – debris slope deposit: sandy silty gravel (Holocene); Fd – Mt. Pettino debris flow and pediment alluvial deposits: dense and poor to well-cemented calcareous gravel with tephra horizons (Upper-Pleistocene); Ta – Vetoio stream terraced alluvium: gravel, sand and clayey-sandy silt with tephra horizons (Upper-Middle Pleistocene); Rs – red soil: reddish clayey silt with gravel (Upper-Middle Pleistocene); Brs – L'Aquila breccia: dense sandy silty gravel, poor to well-cemented calcareous breccia and sand (Middle Pleistocene); FL1 – upper fluvial-lacustrine unit: sand with levels of organic clay and silt, sandy clayey silt with interbedded sand and lignite (Middle-Lower Pleistocene); FL2 – middle fluvial-lacustrine unit: gravel, sand and clay (Lower Pleistocene); FL3 – lower fluvial-lacustrine unit: clay, sand and gravel (Lower Pleistocene-Upper Pliocene). Meso-Cenozoic carbonate unit: Cbr – limestone with bryozoa and lithotamnia (Lower Miocene). 1 – borehole with depth < 30 m; 2 – borehole with depth 30–50 m; 3 – borehole with depth > 50 m; 4 – fountain 99 Cannelle; 5 – geologic cross-section line; 6 – seismic station; 7 – studied area outline. B1,…, B16 are the studied boreholes. Modified after [38].

31

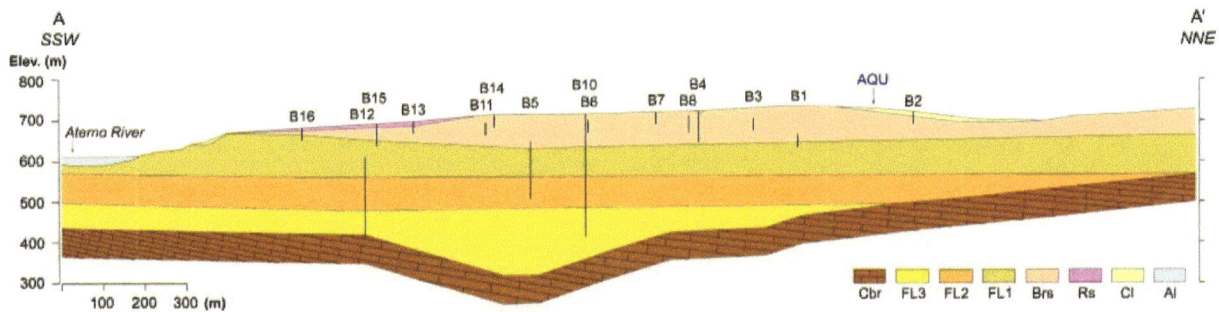

Fig. 3. Geologic cross section of the studied area (its line is positioned in Figure 2). Quaternary units: Al – Aterno river alluvium: coarse- and fine-grained deposits (Holocene); Cl – colluvium: sandy silty gravel, gravelly sandy silt and silty sand (Holocene); Rs – red soil: reddish clayey silt with gravel (Upper-Middle Pleistocene); Brs – L'Aquila breccia: dense sandy silty gravel, poor to well-cemented calcareous breccia and sand (Middle Pleistocene); FL1 – upper fluvial-lacustrine unit: sand with levels of organic clay and silt, sandy clayey silt with interbedded sand and lignite (Middle-Lower Pleistocene); FL2 – middle fluvial-lacustrine unit: gravel, sand and clay (Lower Pleistocene); FL3 – lower fluvial-lacustrine unit: clay, sand and gravel (Lower Pleistocene-Upper Pliocene). Meso-Cenozoic carbonate unit: Cbr – limestone with bryozoa and lithotamnia (Lower Miocene). B1,…, B16 are the studied boreholes and AQU is the seismic station.

The Quaternary deposits, which outcrop in the medieval L'Aquila hill, are represented by L'Aquila breccia (Brs in Fig. 2, 3, 4 and 5) that originated from flash floods, debris flows and rock avalanches, which occurred in the Middle Pleistocene [2]. L'Aquila breccia is generally stratified, reaches a maximum thickness of about 90 m and overlays the Lower-Middle Pleistocene fluvial-lacustrine sequence (FL1, FL2 and FL3 in Fig. 2, 3, 4 and 5), which consists of alternations of silty sand and silty clay that reaches a maximum thickness of about 300 m. The carbonate bedrock (Cbr in Fig. 2, 3, 4 and 5) is made up of Lower Miocene limestone with bryozoa and lithotamnia [1]. The shallow deposits consist of detrital colluvium, red soils and anthropogenic filling materials (Cl, Rs and Fi, respectively, in Fig. 2, 3, 4 and 5), which discontinuously cover, with a maximum thickness of about 10-15 m, the underlying L'Aquila breccia (Fig. 2 and 3). Geotechnical and geophysical data, which are obtained from the borehole logs and in situ seismic tests (Fig. 4 and 5) [1,14,48], allow us to subdivide the investigated lithotypes into nine units and define the 2D subsurface model in the historical City of L'Aquila (Fig. 3). Among the Quaternary deposits, it is possible to recognize eight units (Al, Cl, Fi, Rs, Brs, FL1, FL2 and FL3 in Fig. 2, 3, 4 and 5), which cover the Meso-Cenozoic carbonate bedrock that lays at depths ranging between 150 and 400 m below the ground level, as revealed from gravity and borehole data [48]. About the hydrogeological setting, L'Aquila breccia contains an unconfined aquifer [37] as revealed from the natural spring called Fountain 99 Cannelle, which emerges at an elevation of about 633 m at the contact between the units Brs and FL1 (Fig. 2). The water-table is estimated to lie at an elevation of approximately 670 m within the studied 2D subsurface model (Fig. 3 and 6). The geotechnical units are summarized in Table 1 with the geotechnical and geophysical features that are obtained from laboratory and in situ tests.

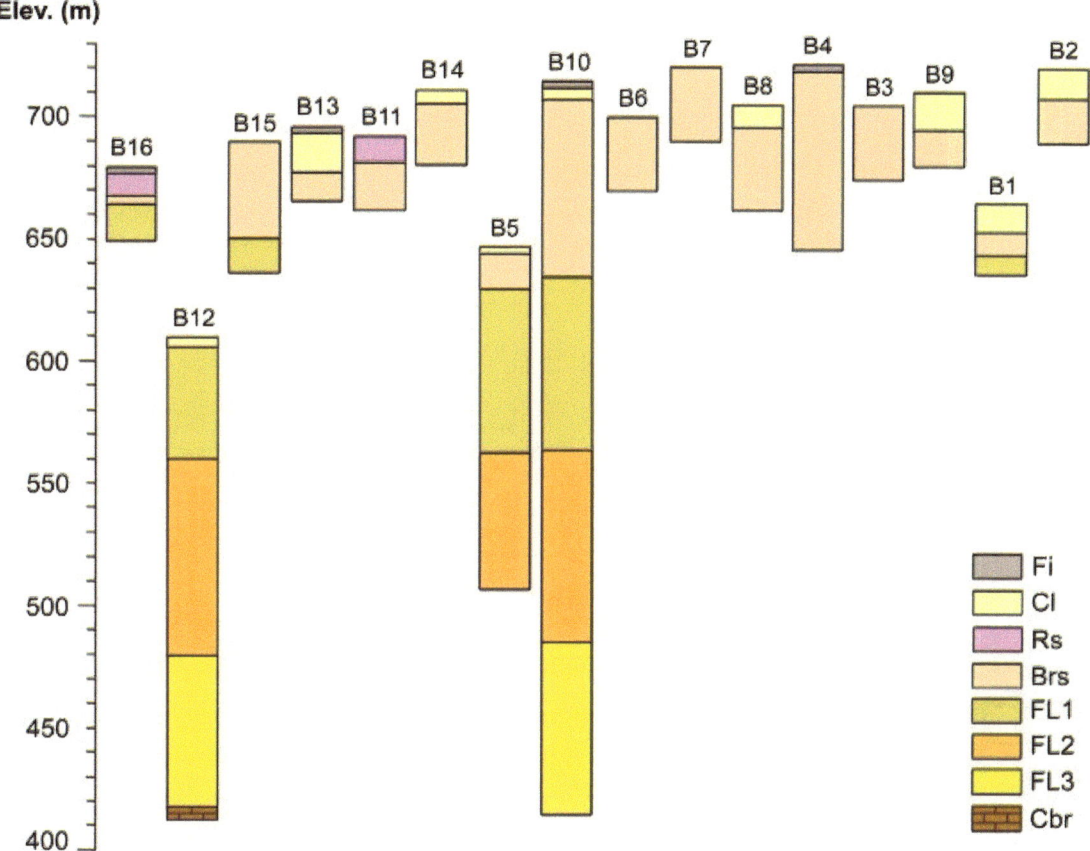

Fig. 4. Borehole logs of the investigated area (their position is in Figure 2). Quaternary units: Fi – filling anthropogenic material (Holocene); Cl – colluvium: sandy silty gravel, gravelly sandy silt and silty sand (Holocene); Rs – red soil: reddish clayey silt with gravel (Upper-Middle Pleistocene); Brs – L'Aquila breccia: dense sandy silty gravel, poor to well-cemented calcareous breccia and sand (Middle Pleistocene); FL1 – upper fluvial-lacustrine unit: sand with levels of organic clay and silt, sandy clayey silt with interbedded sand and lignite (Middle-Lower Pleistocene); FL2 – middle fluvial-lacustrine unit: gravel, sand and clay (Lower Pleistocene); FL3 – lower fluvial-lacustrine unit: clay, sand and gravel (Lower Pleistocene-Upper Pliocene). Meso-Cenozoic carbonate unit: Cbr – limestone with bryozoa and lithotamnia (Lower Miocene).

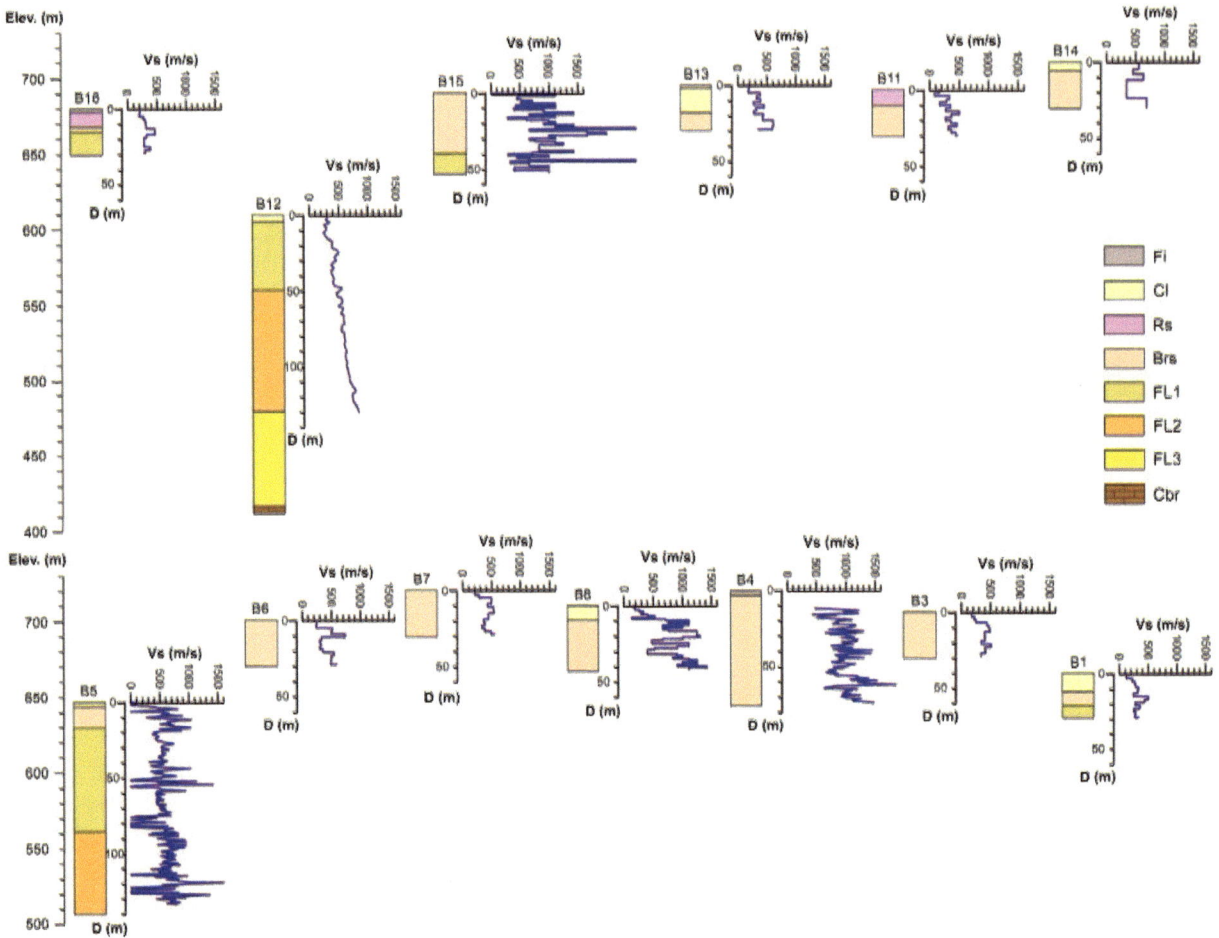

Fig. 5. Borehole logs with the S wave velocity profiles (their position is in Figure 2). Quaternary units: Fi – filling anthropogenic material (Holocene); Cl – colluvium: sandy silty gravel, gravelly sandy silt and silty sand (Holocene); Rs – red soil: reddish clayey silt with gravel (Upper-Middle Pleistocene); Brs – L'Aquila breccia: dense sandy silty gravel, poor to well-cemented calcareous breccia and sand (Middle Pleistocene); FL1 – upper fluvial-lacustrine unit: sand with levels of organic clay and silt, sandy clayey silt with interbedded sand and lignite (Middle-Lower Pleistocene); FL2 – middle fluvial-lacustrine unit: gravel, sand and clay (Lower Pleistocene); FL3 – lower fluvial-lacustrine unit: clay, sand and gravel (Lower Pleistocene-Upper Pliocene). Meso-Cenozoic carbonate unit: Cbr – limestone with bryozoa and lithotamnia (Lower Miocene).

Ground Motion Record Analysis

The strong- and weak-motion data are analyzed for 363 signals, including (the three-component acceleration waveforms), which are recorded using the accelerometer stations AQK and AQU (Table 2) for the events with magnitude ranging from 3.5 to 6.1, mainly during the 2009 L'Aquila seismic sequence [24]. The corrected acceleration time-histories are downloaded from ITACA 2.0 [30, 36] to analyze the observed maximum acceleration (Tables 4 and 5) and to recompute the 5 % damping elastic acceleration response spectra [7,34]. The H/V (horizontal-to-vertical) PGA (peak ground acceleration) ratio of the recorded 2009 events with a magnitude range of 3.0 to 6.1 is computed, for the stations AQU (on colluvium), AQG and T0102 (on rock) (Table 2), then the variation with earthquake magnitude is considered. Finally, the single station

H/V response spectral ratio [9,11,16,23,28,32,39,42,45,46] is computed using the experimental acceleration response spectra to evaluate the local site effects at stations AQK and AQU.

Table 1
Geotechnical and geophysical features of the units. γ – unit weight; c – cohesion; φ – friction angle; V_S – S-wave velocity; V_P – P-wave velocity.

Unit	Lithology	γ (kN/m^3)	c (kPa)	φ (degrees)	V_S (m/s)	V_P (m/s)
Cl	Sandy silty gravel	19	5-10	25-35	250-500	800-1400
Rs	Clayey silt with gravel	19	10-30	25-35	200-500	800-1500
Brs	Sandy silty gravel, breccia and sand	20-22	0-20	30-35	300-1500	1000-2000
FL1	Sand with levels of organic clay and silt. Sandy clayey silt with interbedded sand and lignite	19	10-40	24-34	300-500	900-1500
FL2	Gravel, sand and clay	19	10-40	24-34	400-800	1200-2400
FL3	Clay, sand and gravel	19	10-40	24-34	800-1600	2400-4800
Cbr	Limestone	24	250-350	30-40	1750-1900	5200-5700

Table 2
Characteristics of the seismic stations.

Station	Latitude	Longitude	Lithology	V_{S30} (m/s)	EC8 Class	Installation Date	Removal Date
AQG	42.37347	13.33703	Limestone	685	B	1997-01-01	–
T0102	42.39670	13.31390	Limestone	–	B*	2009-04-06	2009-09-01
AQK	42.34497	13.40095	Breccia/Sand	717	B	2005-12-02	–
AQU	42.35388	13.40193	Colluvium	–	C*	2008-02-18	–
AQA	42.37553	13.33930	Alluvium	552	B	2001-04-17	–
AQV	42.37722	13.34389	Alluvium	474	B	1997-01-01	–

Dynamic Modeling

To identify the seismic response of the horizontally layered soil system, which is shown in the geologic cross-section (Fig. 3), dynamic analyses are carried out on a 80-meter wide and 240-meter deep soil column at station AQU (Fig. 6), using QUAKE/W that is a software based on finite element formulations with a direct integration scheme in the time domain [40]. The mesh pattern of model consists of quadrangular and triangular elements (with global size of 10 m). The initial static analyses are performed to determine the initial state of stress in the ground before starting the dynamic analyses. The initial pore-water pressure conditions are computed by specifying the water table, which is estimated at the elevation of 670 m (Fig. 6).

The lower boundary [8], where the movement is fixed in both the x and y directions, is positioned on the interface between the lower fluvial-lacustrine unit (FL3) and the Meso-Cenozoic carbonate bedrock (Cbr). Moreover, the vertical boundaries are kept fixed in order to prevent energy and numerical dispersion (Fig. 6) [25].

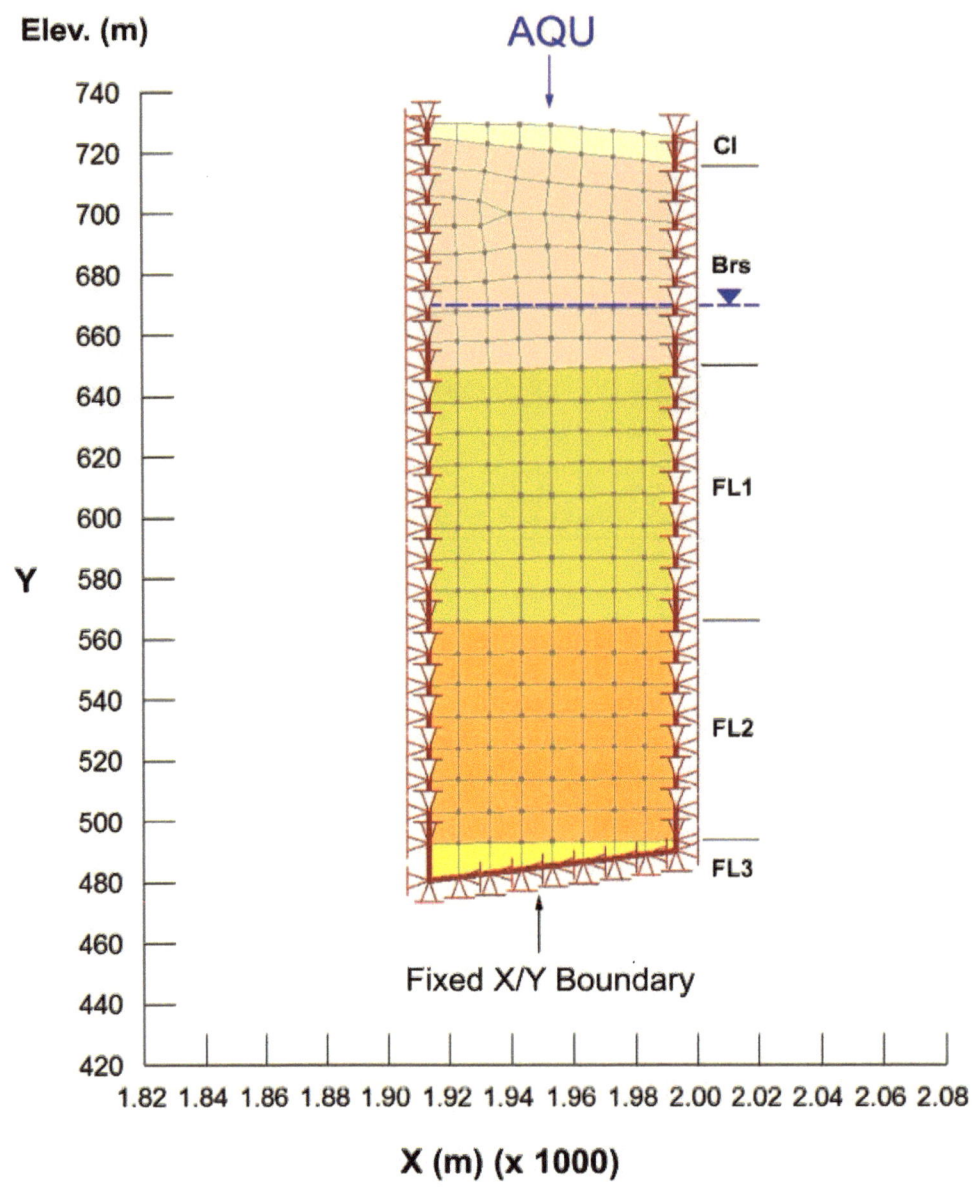

Fig. 6. Model setup of the material zones (Cl, Brs, FL1, FL2 and FL3 in Figures 3, 4 and 5), the water-table, the boundary condition and the mesh pattern that are used to perform the dynamic modeling. AQU is the seismic station and the blue dashed line denotes the water table.

To simulate the nonlinear effective stress behavior [3,10,12,21,27,43,49,50], 202 nonlinear dynamic analyses based on the MFS (Martin-Finn-Seed) pore-water pressure model [5,31,51] are executed. The soil geotechnical and dynamic properties that are used as input parameters in the analyses are presented in Table 3. Besides, the shear modulus G_{max} is defined as a function of the effective vertical stress (Fig. 7) taking into account the S-wave velocity profiles that are displayed in Figure 5. The rebound modulus E_r [31] and the MFS pore-water pressure functions [5,51] that are estimated for different materials are shown in Fig. 8.

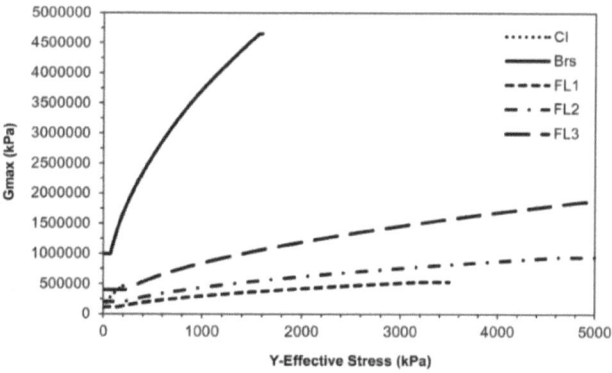

Fig. 7. Variation of the shear modulus G_{max} of modeled materials with vertical (y) effective stress.

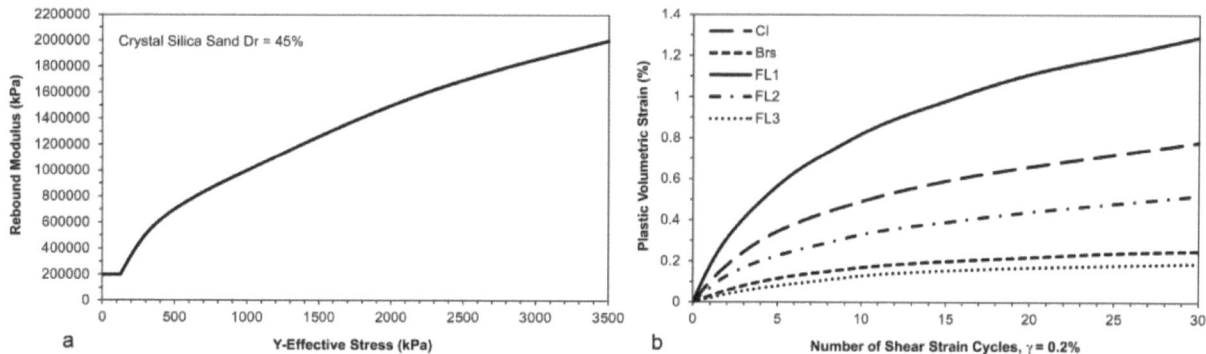

Fig. 8. (a) Variation of the rebound modulus with vertical (y) effective stress, which is obtained for silica sand at a relative density of 45% (Martin et al., 1975); (b) Estimated variation of the plastic volumetric strain of modeled materials with a number of shear strain cycles.

The horizontal and vertical acceleration time-histories, which are recorded by stations AQK and AQG mainly during the 2009 events (Tables 4 and 5), are used as the input motions at the lower boundary of soil column. In particular, the imported records are scaled down by the factors 2 and 3 in order to relate them to the motion at the soil-rock interface, where rock is overlain by a stratum of soil. In fact, when seismic waves travel upwards from deep in the ground they are refracted and reflected at the soil-rock interface, which may alter the motion at the soil-rock interface and consequently the motion is different than at the outcrop locations. Thus, the ground motions recorded by stations AQG and AQK are purposely scaled and used as the input motions to evaluate the seismic response of soil profile under different ground motions excitation [8,15].

To simulate the seismic response of Quaternary sediments under the 2009 M_w 6.1 main shock excitation, the signals recorded by AQG, which is the only near-fault station on rock that registered that event, are used as the input motions. In particular, during each analysis the model is simultaneously subjected to the horizontal and vertical accelerations of the earthquake. Finally, the computed acceleration response spectra are normalized to calculate the numerical single station H/V response spectral ratio.

Table 3
Geotechnical and dynamic properties of the units used as input parameters in the numerical simulations.

Unit	Cl	Brs	FL1	FL2	FL3
Unit weight (kN/m^3)	19	21	19	19	19
Cohesion (kPa)	10	10	10	20	40
Friction angle (degrees)	26	32	24	29	34
Poisson's ratio	0.40	0.46	0.47	0.48	0.49
Damping ratio	0.05	0.05	0.05	0.05	0.05
Max Damping ratio	0.2	0.1	0.4	0.3	0.2

Table 4
Characteristics of the analyzed seismic events that are recorded by seismic stations AQK and AQU (on Quaternary sediment). The seismic events recorded by station AQK and marked with an asterisk are also used as the input motions in numerical simulations. H-PGA denotes the horizontal peak ground acceleration (geometric mean of north–south and east–west components).

Date_Time (UTC)		M_W	M_L	AQK			AQU		
				Epicentral distance (km)		H-PGA (cm/s^2)	Epicentral distance (km)		H-PGA (cm/s^2)
20090406_01:32:40	(Main shock)	6.1	5.9	1.8		335.06	2.4		277.15
20090407_21:34:30*		4.5	4.3	3.7		122.18	3.1		50.01
20090407_17:47:37*		5.5	5.4	8.4		84.30	8.9		65.62
20090330_13:38:39		4.4	4.1	3.4		75.84	4.2		47.69
20090410_15:46:18*		–	3.5	1.4		71.20	–		–
20090412_09:48:59*		–	3.5	1.5		71.20	–		–
20090406_02:37:04*		5.1	4.6	6.2		62.10	6.1		31.65
20090406_03:56:46*		4.5	4.1	1.7		57.38	2.5		28.16
20090407_09:26:29*		5.1	4.8	1.5		54.72	2.3		41.34
20090409_00:53:00*		5.4	5.1	16.5		44.14	15.6		31.69
20090406_23:15:37*		5.1	5.0	13.2		37.16	12.2		31.12
20090405_20:48:54		4.2	3.9	2.7		36.61	3.6		13.57
20090712_22:14:25*		–	3.8	1.1		35.94	–		–
20090406_16:38:10*		4.4	4.1	5.5		28.67	5.3		19.39
20090712_08:38:51*		4.3	4.2	2.6		26.89	3.4		13.18
20090622_20:58:40*		4.7	4.6	12.7		25.83	11.8		8.70
20090406_01:40:51		–	4.1	8.0		23.68	7.0		3.32
20090406_01:41:33		–	4.0	7.6		23.66	7.3		19.73
20090406_10:36:18		–	3.5	0.8		23.03	–		–
20090409_19:38:17*		5.2	5.0	18.2		22.76	17.2		17.15
20090413_19:09:49*		–	3.8	4.6		21.65	4.6		16.30
20090407_09:24:57*		–	3.7	0.8		21.66	–		–
20090406_04:47:54*		–	3.6	3.9		21.05	–		–
20090406_02:27:46		–	3.9	7.1		20.16	–		–
20090530_02:55:10*		–	3.6	5.1		19.45	–		–
20090409_13:19:34*		–	4.1	11.7		18.64	11.9		6.30
20090703_11:03:08*		4.1	4.1	5.8		18.42	4.9		15.25
20090413_21:14:24*		5.0	5.0	17.1		17.87	16.1		13.97
20090430_13:01:02*		–	3.5	3.4		16.64	–		–
20090406_01:36:29		–	4.7	–		–	3.2		31.90
20090406_01:42:50		–	4.2	5.5		14.91	6.4		14.11
20090409_03:14:52*		4.4	4.6	4.2		11.22	4.5		7.61
20090409_04:32:45*		4.3	4.2	12.9		14.99	11.9		7.98
20090418_09:05:56*		–	4.0	11.7		14.65	10.8		5.12
20090924_16:14:58*		4.2	4.1	13.4		12.55	–		–
20090330_13:43:27		–	3.4	3.8		13.50	–		–
20090405_22:39:42		–	3.5	2.2		15.90	–		–
20090406_01:38:47		–	3.6	6.5		10.89	–		–
20090406_10:12:36		–	3.6	3.4		10.68	–		–
20090406_21:56:54		–	3.9	6.5		13.55	–		–

Date_Time (UTC)	M_W	M_L	AQG Epicentral distance (km)	AQG H-PGA (cm/s²)	T0102 Epicentral distance (km)	T0102 H-PGA (cm/s²)
20090407_09:30:57	–	3.8	1.5	14.10	–	–
20090413_19:17:58*	–	3.5	5.0	13.55	–	–
20090514_20:30:55*	–	3.5	6.6	14.17	–	–
20090703_01:14:07*	–	3.6	1.9	14.97	–	–
20090408_22:56:50	4.1	4.2	17.1	7.50	16.2	5.47
20090406_01:44:35	–	3.7	5.6	9.00	–	–
20090406_22:47:14	–	3.7	8.3	9.63	–	–
20090409_04:43:10	–	4.0	18.2	2.14	17.2	1.31
20090409_22:40:06	–	3.8	17.4	2.36	–	–
20090413_13:36:05	–	3.6	12.0	4.90	–	–
20090414_13:56:21	–	4.0	–	–	22.3	1.16
20090414_17:27:31*	–	3.6	21.6	2.55	–	–
20090414_20:17:27*	4.0	4.1	23.2	1.93	22.3	1.08
20090415_22:53:08*	4.1	4.0	19.8	2.39	18.9	1.56
20090416_17:49:30*	–	4.1	23.8	2.76	22.9	2.15
20090421_15:44:36*	–	3.7	3.4	9.80	–	–
20090423_15:14:08*	4.1	4.0	12.9	6.24	13.7	4.11
20090423_21:49:01*	4.3	4.2	14.8	3.62	15.6	2.99
20090501_05:12:52*	–	3.8	8.2	3.84	–	–
20090509_09:09:40*	–	3.5	2.8	7.65	–	–
20090511_16:59:04*	–	3.5	16.1	1.53	–	–
20090514_06:30:22*	–	3.5	15.3	3.25	–	–
20090514_06:32:34*	–	3.5	15.9	1.92	–	–
20090607_19:21:34*	–	3.6	6.2	4.26	–	–
20090623_00:41:56*	–	4.0	12.3	9.95	11.3	5.48
20090623_08:35:09*	–	3.6	14.0	2.75	–	–
20090703_09:43:54*	–	3.6	3.2	6.98	–	–
20090731_11:05:40*	–	3.8	12.2	3.47	–	–
20090806_15:36:44	–	4.2	–	–	81.5	0.17
20091020_05:07:31*	–	3.6	13.8	2.56	–	–
20100110_08:33:36*	–	4.0	86.1	0.91	85.1	0.67
20100122_12:30:51*	–	3.5	15.3	1.33	–	–
20100415_01:47:36	4.1	3.8	–	–	147.2	0.02
20100828_07:08:03	4.1	4.1	–	–	81.3	0.12
20100831_07:12:52*	–	3.6	23.8	1.25	–	–
20100917_12:20:18	4.3	4.5	–	–	208.7	0.03
20121205_01:18:19	4.0	–	–	–	65.4	0.32
20120315_03:29:05	4.0	4.0	–	–	51.9	0.27
20130216_21:16:09	4.9	4.8	–	–	72.5	0.56
20130721_01:32:34	5.1	4.9	–	–	129.3	0.40
20130822_06:44:50	4.4	4.4	–	–	140.4	0.08
20131229_17:08:43	–	4.9	–	–	139.5	0.25
19971014_15:23:09	5.6	5.5	75.5	8.14	–	–
19970926_09:40:25	6.0	5.8	88.1	4.06	–	–
19970926_00:33:12	5.7	5.7	86.1	5.24	–	–
20071021_03:55:36	4.2	3.9	36.6	1.90	–	–
20080319_14:38:58	4.2	4.1	–	–	209.3	0.03

Table 5

Characteristics of the analyzed seismic events that are recorded by seismic stations AQG and T0102 (on Meso-Cenozoic carbonate rock). The seismic events recorded by station AQG and marked with an asterisk are also used as the input motions in numerical simulations. H-PGA denotes the horizontal peak ground acceleration (geometric mean of north–south and east–west components).

Date_Time (UTC)		M_W	M_L	AQG Epicentral distance (km)		H-PGA (cm/s²)	T0102 Epicentral distance (km)		H-PGA (cm/s²)
20090406_01:32:40*	(Main shock)	6.1	5.9	5.1		457.87	–		–
20090407_17:47:37*		5.5	5.4	14.6		117.71	–		–
20090407_21:34:30*		4.5	4.3	3.0		93.67	–		–
20090406_02:37:04*		5.1	4.6	1.7		75.41	–		–

20090409_00:53:00*	5.4	5.1	12.9	64.91	–	–
20090406_03:56:46*	4.5	4.1	5.9	58.80	–	–
20090407_09:26:29*	5.1	4.8	5.9	55.80	–	–
20090413_21:14:24*	5.0	5.0	14.2	40.20	–	–
20090430_13:01:02*	–	3.5	2.8	39.71	–	–
20090406_04:47:54*	–	3.6	2.5	34.12	–	–
20090622_20:58:40	4.7	4.6	9.1	32.34	7.2	43.95
20090409_19:38:17*	5.2	5.0	14.5	31.93	–	–
20090406_02:27:46	–	3.9	1.7	30.64	–	–
20090409_13:19:34*	–	4.1	7.4	29.24	–	–
20090410_15:46:18*	–	3.5	4.9	25.24	–	–
20090418_09:05:56*	–	4.0	8.3	22.97	–	–
20090413_19:09:49*	–	3.8	2.6	22.34	–	–
20090514_20:30:55*	–	3.5	0.7	22.05	–	–
20090530_02:55:10*	–	3.6	2.6	21.18	5.6	16.34
20090413_19:17:58*	–	3.5	2.4	20.38	–	–
20090409_03:14:52*	4.4	4.6	10.2	18.78	–	–
20090409_04:32:45*	4.3	4.2	12.6	18.76	–	–
20090407_09:24:57*	–	3.7	5.6	18.03	–	–
20090623_00:41:56*	–	4.0	9.0	15.00	7.4	20.75
20090412_09:48:59*	–	3.5	4.7	13.54	–	–
20090410_03:22:22*	3.9	3.9	12.1	12.53	–	–
20090703_01:14:07*	–	3.6	5.4	12.09	–	–
20090509_09:09:40*	–	3.5	3.4	11.49	–	–
20090407_09:30:57*	–	3.8	5.4	10.42	–	–
20090712_08:38:51	4.3	4.2	–	–	9.3	15.74
20090703_11:03:08	4.1	4.1	–	–	6.3	14.77
20090406_01:44:35*	–	3.7	8.6	4.23	–	–
20090409_04:43:10*	–	4.0	15.3	3.41	–	–
20090409_22:40:06*	–	3.8	12.8	6.38	–	–
20090413_13:36:05*	–	3.6	12.5	4.63	–	–
20090414_13:56:21*	–	4.0	19.3	3.68	–	–
20090414_17:27:31*	–	3.6	17.3	2.63	–	–
20090414_20:17:27*	4.0	4.1	19.0	5.57	–	–
20090415_22:53:08*	4.1	4.0	15.7	3.55	–	–
20090416_17:49:30*	–	4.1	19.3	7.79	–	–
20090421_15:44:36*	–	3.7	6.2	5.37	–	–
20090423_15:14:08*	4.1	4.0	18.6	8.49	–	–
20090423_21:49:01*	4.3	4.2	20.3	5.20	–	–
20090501_05:12:52*	–	3.8	14.1	3.16	–	–
20090511_16:59:04*	–	3.5	13.2	2.23	–	–
20090514_06:30:22*	–	3.5	13.1	6.87	–	–
20090514_06:32:34*	–	3.5	13.9	5.24	–	–
20090523_17:24:22	–	3.2	–	–	12.1	1.28
20090525_16:37:28	–	3.4	–	–	25.2	0.33
20090531_03:36:19	–	3.0	–	–	6.2	1.36
20090531_09:09:50	–	3.1	–	–	24.6	0.16
20090604_12:40:32	–	3.1	–	–	25.3	0.22
20090607_19:21:34*	–	3.6	12.1	1.82	14.9	1.08
20090611_07:04:50	–	3.3	–	–	12.6	0.34
20090616_18:52:58	–	3.1	–	–	13.7	1.00
20090619_19:47:10	–	3.2	–	–	12.3	2.30
20090623_08:35:09*	–	3.6	10.4	3.67	–	–
20090626_07:14:30*	–	3.5	24.2	1.00	–	–
20090731_11:05:40*	–	3.8	18.2	5.22	–	–
20090806_15:36:44	–	4.2	–	–	88.2	0.15
20100122_12:30:51*	–	3.5	21.2	2.53	–	–
20100831_07:12:52*	–	3.6	18.3	2.01	–	–

Peak Ground Acceleration

Table 6 presents a comparison between the observed and computed peak ground acceleration for stations AQK, AQU (in medieval L'Aquila City), AQV, AQA and AQG [41] (in Upper Aterno River Valley) (Table 2). The values are notably consistent for the horizontal component of ground motion and reveal a maximum acceleration of about 0.6g at station AQV, which is double the value at stations AQK and AQU. The higher value at station AQV may be related to the sharp seismic wave velocity contrast between the alluvial deposits and the carbonate bedrock, which occurs at a shallow depth of about 50 m [41], whereas the considerable thickness (200-400 m) of dissipative sediments that cover the carbonate bedrock (Fig. 3, 4 and 6) reduces the seismic ground motion in the historical city of L'Aquila. Except for station AQG, the computed vertical peak acceleration underestimates the observed value, which suggests that some near-source effects, such as conversion of phases or vertical incidence angle [4,29], amplify the vertical acceleration in Upper Aterno River Valley and historic L'Aquila City center.

Table 6
Observed and computed maximum acceleration at seismic stations AQK, AQU, AQV, AQA and AQG for the 2009 $M_W = 6.1$ L'Aquila earthquake.

Seismic station	Horizontal PGA (g)		Vertical PGA (g)	
	Observed	Computed	Observed	Computed
AQK	0.34	–	0.36	–
AQU	0.28	0.29	0.31	0.24
AQV	0.60	0.59*	0.50	0.32*
AQA	0.42	0.45*	0.44	0.23*
AQG	0.47	0.49*	0.24	0.23*

* = data from [41]

This hypothesis is corroborated from Figure 9 showing the variation of H/V PGA ratio with earthquake magnitude normalized to epicentral distance (Km). In particular, the observed H/V peak acceleration ratio at stations AQG and T0102 (on rock outcrops) exhibits a mean value near 2 that does not notably vary when the earthquake magnitude increases (Fig. 9a), whereas at station AQU (on colluvium) (Fig. 2, 3 and 6) it distinctly displays a downward trend (Fig. 9b).

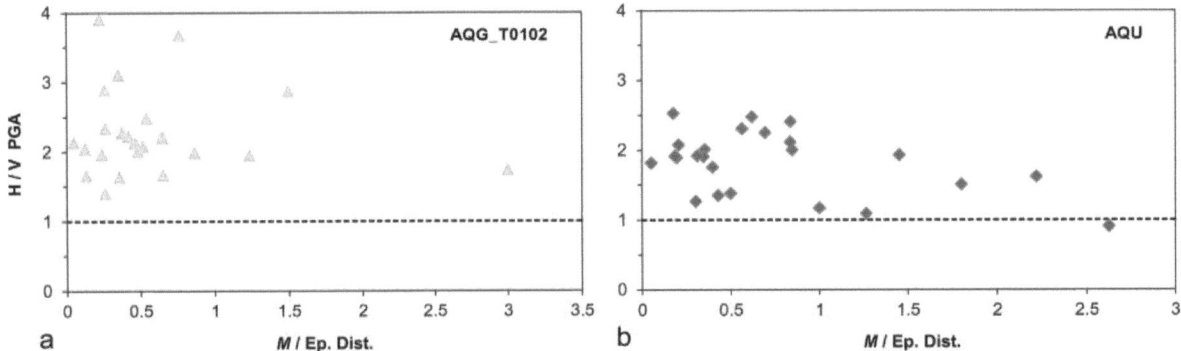

Fig. 9. Variation of the horizontal to vertical peak ground acceleration ratio (H/V PGA) with ratio of earthquake magnitude to epicentral distance (M/Ep. Dist.): (a) the variation at seismic stations AQG and T0102 (on Meso-Cenozoic carbonate rock); (b) the variation at seismic station AQU (on Quaternary colluvium).

Observed vs. Numerical H/V Response Spectral Ratio

The H/V response spectral ratios from the ground motion records analysis (observed in Fig. 10 and 11) and the nonlinear dynamic modeling (computed in Fig. 11), provided a better understanding of the soil dynamic behavior at seismic stations AQK and AQU.

At AQK that is located by the contact between geologic units Brs and FL1 (Fig. 1 and 2), for all the analyzed seismic events ($10 > $ PGA $ > 100$ cm/s^2), the observed H/V spectral ratio exhibits a maximum amplification at 0.5-0.6 Hz (f_0) (Fig. 10a, b and c), which may be related to the superposition of Quaternary sediments on the Meso-Cenozoic carbonate bedrock at a depth of 200-400 m (Fig. 3, 4 and 5) [11,19,22,23,26,28,33,41]. The level of maximum amplification increases from about 2, which is the mean level for the events with PGA < 100 cm/s^2, to 4-5 during the 2009 M_w 6.1 L'Aquila main shock (Fig. 10c), probably because the Quaternary deposits are excited by seismic phases that are characterized by a very high energy at 0.5-0.6 Hz and by a peculiar wave-field incidence angle [25].

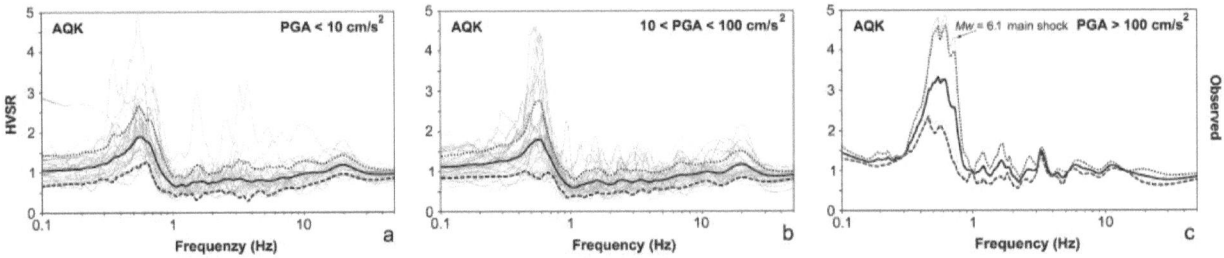

Fig. 10. Observed (a, b, c) horizontal to vertical response spectral ratio (HVSR) for three levels of peak ground acceleration (PGA) at seismic station AQK. The black solid curves show the mean values and the black dashed curves show the mean values +/- 1 standard deviation.

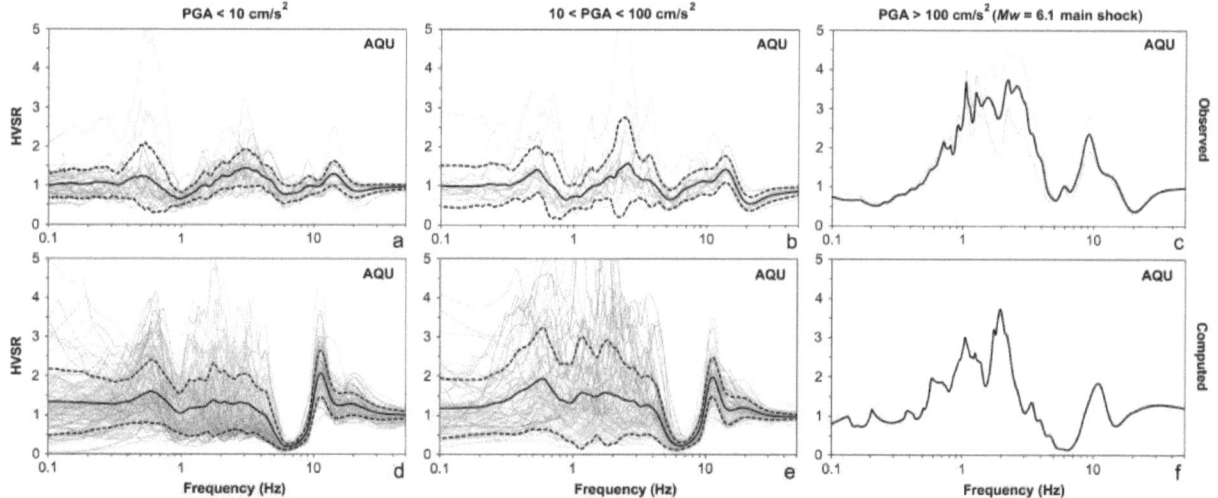

Fig. 11. Observed (a, b, c) and computed (d, e, f) horizontal to vertical response spectral ratio (HVSR) for three levels of peak ground acceleration (PGA) at seismic station AQU. The black solid curves show the mean values and the black dashed curves show the mean values +/- 1 standard deviation.

At AQU that is located by the top of L'Aquila Hill (Fig. 1, 2 and 3), for the weak seismic events (PGA < 10 cm/s²), the observed H/V spectral ratio displays maximum amplification at 0.5 Hz, 3 Hz and 14 Hz (Fig. 11a). Similarly to AQK, the 0.5 Hz (f_0) amplification may be related to the superposition of Quaternary deposits on the Meso-Cenozoic carbonate bedrock at a depth of 200-300 meters (Fig. 3, 4 and 5). Besides, the 3 Hz (f_1) amplification may be associated to the strong impedance ratio [25,41] that occurs when L'Aquila breccia unit Brs contacts upper fluvial-lacustrine unit FL1 (Fig. 3, 4 and 5) and the 14 Hz (f_2) amplification may be related to the shallow seismic wave velocity contrast between Quaternary units Cl and Brs (Fig. 3, 4 and 5).

For the stronger seismic events (10 < PGA < 100 cm/s²), the observed H/V spectral ratio shows maximum amplification at 0.5 Hz and 14 Hz, while the 3 Hz (f_1) predominant frequency that is observed for the weaker events shifts to 2.5 Hz (Fig. 11b).

For the 2009 M_w 6.1 main shock, the observed H/V spectral ratio clearly reveals maximum amplification at 0.6-0.7 Hz, 1-2 Hz and 10 Hz, which denotes a distinct reduction of the predominant frequencies f_1 and f_2. In addition, the mean level of maximum amplification increases to 3-4 (Fig. 11c).

The simultaneous decrease of the resonance frequency f_1 and the increase of the maximum amplification level suggest that somewhere the behavior of Quaternary sediments filling L'Aquila Basin is probably nonlinear. This hypothesis is validated by the computed H/V spectral ratio (Fig. 11d, e and f), which shows frequencies, levels and variation of maximum amplification that are, for the most part, consistent with those revealed from the observed H/V spectral ratio. In particular, the frequency f_1 shifts from about 3 Hz to 1-2 Hz when the earthquake magnitude increases for both the observed and computed H/V spectral ratios, whereas the decrease of frequency f_2 (from about 14 Hz to 9 Hz) is only revealed from the observed H/V spectral ratio, which probably denotes that a peculiar incidence angle of the seismic waves is the main cause of observed reduction [25].

Excess Pore-Water Pressure and Nonlinear Behavior

The nonlinear dynamic analyses based on the MFS pore-water pressure model [31] provided insights on the excess PWP (pore-water pressure) that was generated under seismic loading in the subsurface at seismic station AQU. In particular, the simulation of 2009 M_w 6.1 main shock reveals a maximum excess pore-water pressure of about 380 kPa in the upper fluvial-lacustrine unit FL1 (Fig. 12d), which softens and amplifies the seismic motion (Fig. 12a, b and c). Figures 13a and b, respectively, show the variation of excess PWP with elevation for 200 time steps of 10 second analysis and the time-history of excess pore-water pressure for different elevations below seismic station AQU. The excess pore-water pressure, which reaches maximum values ranging between 50 kPa (in the unit FL2) and 350 kPa (in the unit FL1) after 3 to 7 seconds of shaking (Fig. 13b), leads to a reduction of the vertical effective stress and shear modulus (Fig. 14a and b). In particular, the vertical effective stress and shear modulus, which are computed at the interface between units Brs and FL1, respectively reduce up to 80 % and 60 % of their initial values after about 3 seconds of shaking, which is compatible with the variation of fundamental frequency f_1 from 3 Hz to 1-2 Hz (Fig. 11). In addition, the graphs XY-shear stress vs. XY-shear strain (Fig. 15a, b, c and d), which are computed for different elevations below seismic station AQU, validate the nonlinear seismic response due to the generation of excess pore-water pressure. In particular, Fig. 15c confirms the strong nonlinear behavior occurring at the interface between L'Aquila breccia unit Brs and upper fluvial-lacustrine unit FL1.

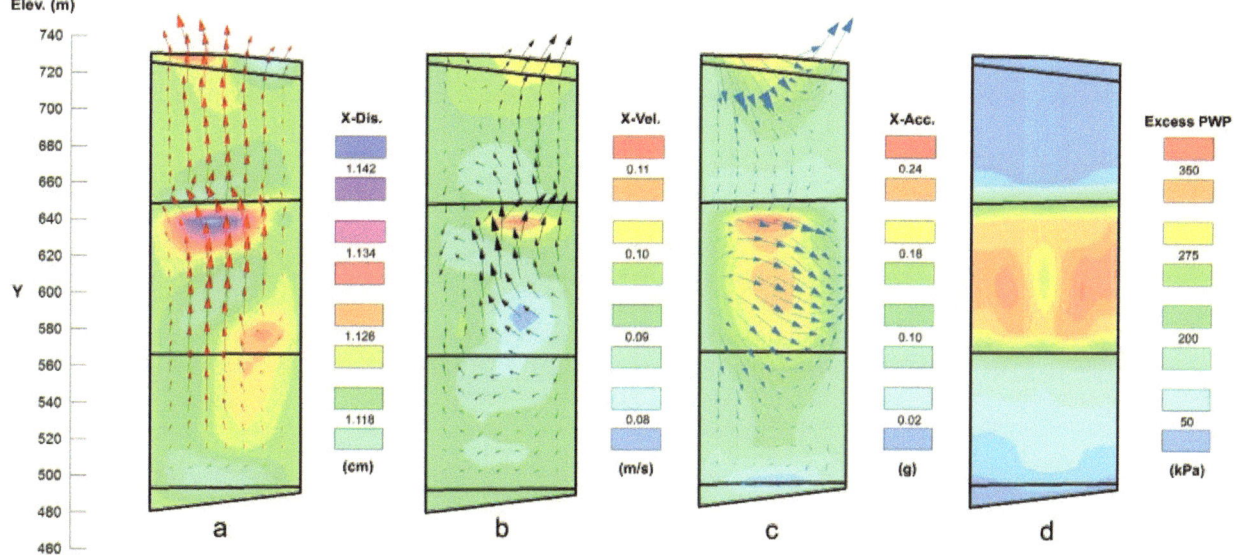

Fig. 12. Numerical simulation of 2009 M_W 6.1 L'Aquila earthquake at seismic station AQU: (a) X-displacement (X-Dis.) contour map and the relative displacement vectors (red arrows) at 2.3 seconds of the nonlinear dynamic analysis; vector length magnification: 44000 times; (b) X-velocity (X-Vel.) contour map and the relative velocity vectors (black arrows) at 3.5 seconds of the nonlinear dynamic analysis; vector length magnification: 674 times; (c) X-acceleration (X-Acc.) contour map and the relative acceleration vectors (blue arrows) at 2.3 seconds of the nonlinear dynamic analysis; vector length magnification: 131 times; (d) excess pore-water pressure (PWP) contour map at 7 seconds of the nonlinear dynamic analysis.

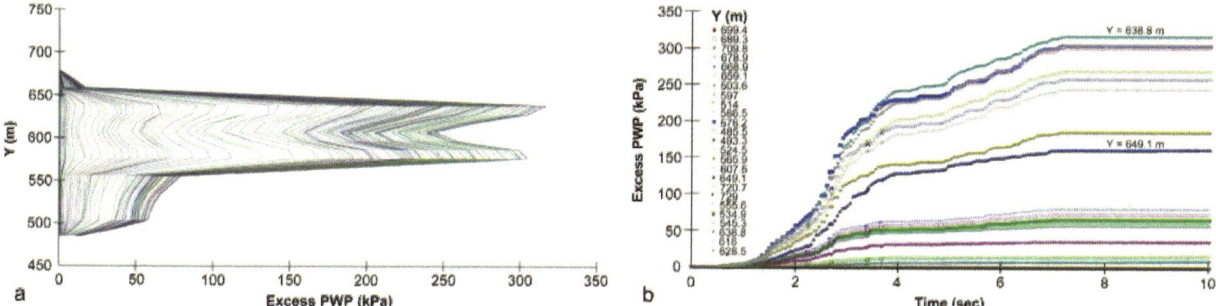

Fig. 13. Numerical simulation of 2009 M_W 6.1 L'Aquila earthquake at seismic station AQU: (a) excess pore-water pressure (PWP) vs. elevation (Y) for 200 time steps of ten-second dynamic analysis ; (b) time-histories of the excess pore-water pressure that are computed for different elevations.

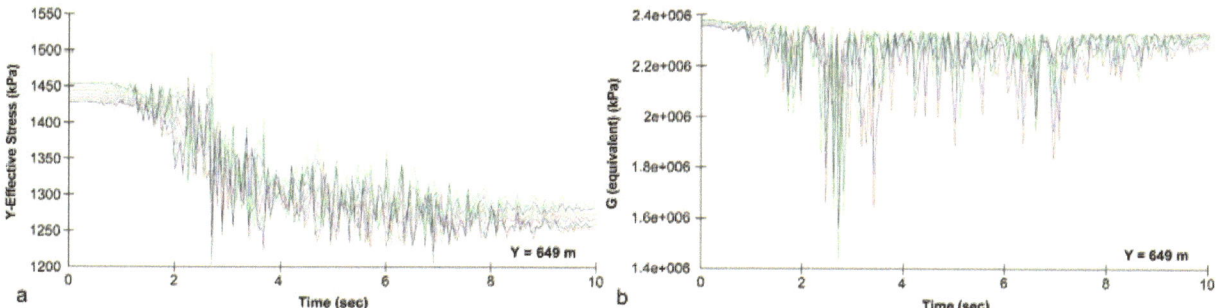

Fig. 14. Numerical simulation of 2009 M_W 6.1 L'Aquila earthquake at seismic station AQU: time histories of the y-effective stress (a) and the shear modulus G (equivalent) (b) that are computed at the interface between units Brs and FL1. Y denotes the elevation.

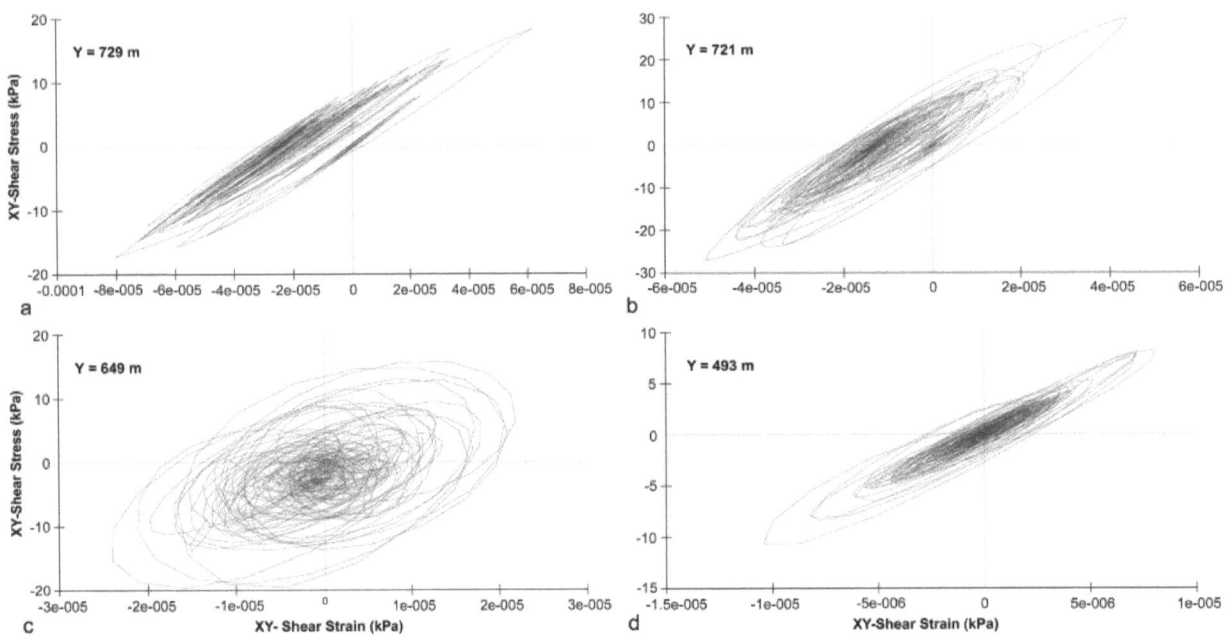

Fig. 15. Numerical simulation of 2009 M_W 6.1 L'Aquila earthquake at seismic station AQU:
XY-shear stress vs. XY-shear strain graphs, which are computed at the elevations of 729 m (a), 721 m (b), 649 m (c) and 493 m (d). Y denotes the elevation.

Shear Stress-Strain State and Seismic Motion

The numerical simulation of 2009 M_w 6.1 L'Aquila earthquake allows us to understand the variations of shear stress-strain state and seismic motion within the soil profile below seismic station AQU. Fig. 16 shows that lower shear stresses and higher shear strains occur at the shallow colluvial unit Cl, at the upper fluvial-lacustrine unit FL1 that is located between elevations 560 and 650 m and at the bottom of fluvial-lacustrine sequence, whereas the higher dynamic shear stresses and lower shear strains occur in the stiff L'Aquila breccia unit Brs (Fig. 16c and d).

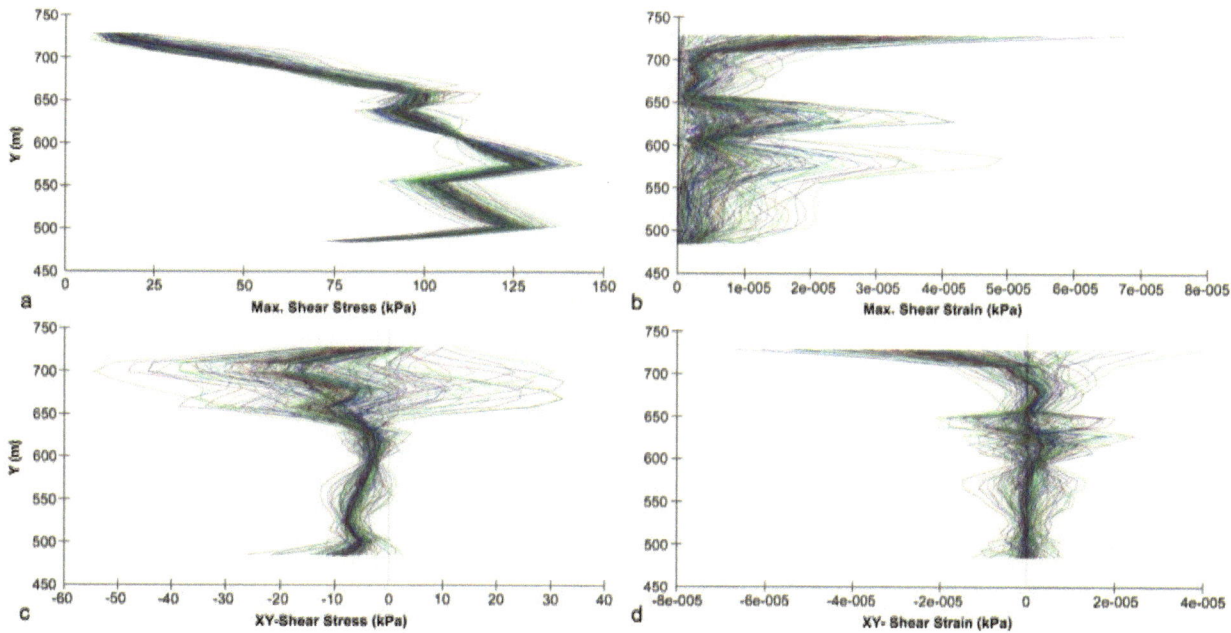

Fig. 16. Numerical simulation of 2009 M_W 6.1 L'Aquila earthquake at seismic station AQU: (a) Max shear stress, (b) Max shear strain, (c) XY-shear stress and (d) XY-shear strain vs. elevation for 200 time steps of ten-second dynamic analysis. Y denotes the elevation.

The resulting seismic motion amplification (Fig. 17) that happens at the shallow colluvium, at the top and base of the fluvial-lacustrine sequence probably produces the amplification at about 9-14 Hz (f_2), 1-3 Hz (f_1) and 0.5-0.7 Hz (f_0), respectively. In particular, the seismic motion amplification occurring at the upper fluvial-lacustrine unit FL1 probably excited the overlying L'Aquila breccia (Brs) (Fig. 17a, b, c and d), which produced amplification at the frequency band of 1-2 Hz during the 2009 M_w 6.1 L'Aquila earthquake (Fig. 11).

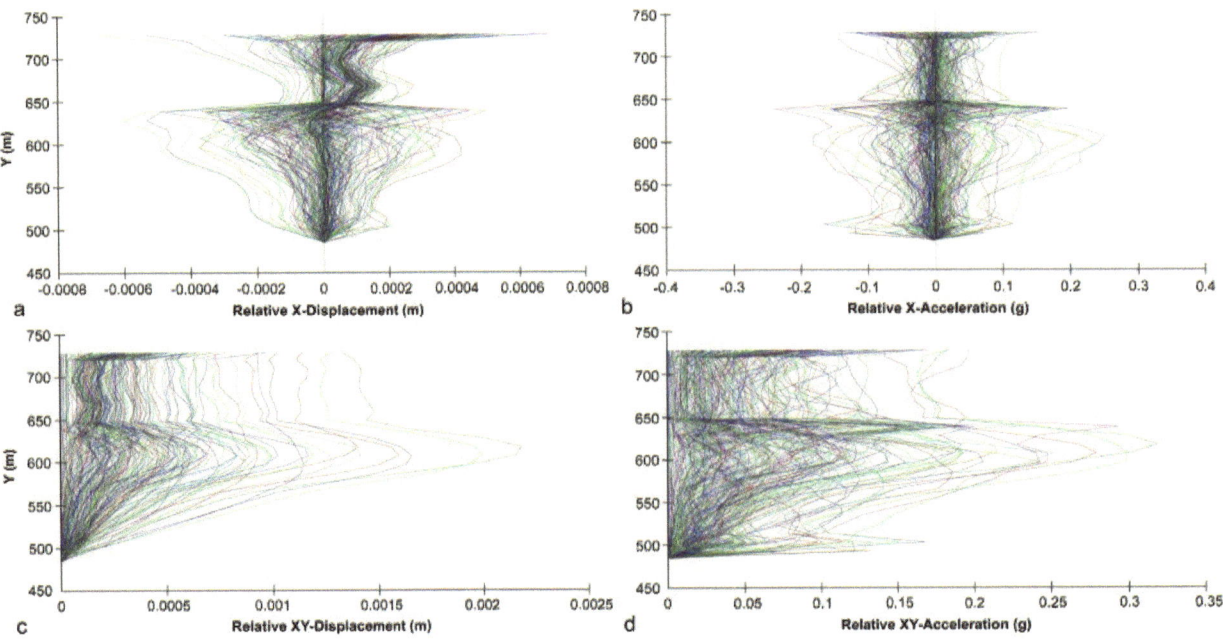

Fig. 17. Numerical simulation of 2009 M_W 6.1 L'Aquila earthquake at seismic station AQU: (a) Relative X-displacement, (b) Relative X-acceleration, (c) Relative XY-displacement and (d) Relative XY-acceleration vs. elevation for 200 time steps of ten-second dynamic analysis. Y denotes the elevation.

Conclusions

The seismic response of deep Quaternary sediments in historical center of L'Aquila City (central Italy) is studied by analyzing 363 seismic signals, which are recorded by the near-fault stations AQK and AQU, and 202 finite element numerical simulations. The signals are mostly recorded during the 2009 L'Aquila seismic sequence and the numerical simulations are performed for the broadband and accelerometer station AQU on a horizontally layered soil column, which is drawn from a geologic cross-section that passes through L'Aquila City center.

The weak- and strong-motion records analysis reveals the following:

- During the 2009 M_w 6.1 main shock stations AQK and AQU (in L'Aquila City center) show a maximum peak ground acceleration that is half the value recorded at station AQV (in Upper Aterno River Valley), which is consistent with the great thickness (200-400 m) of dissipative sediments filling L'Aquila Basin by the city center.

- The H/V peak ground acceleration ratio decreases when the earthquake magnitude increases, which reveals amplification of the vertical component of ground motion.

48

- The H/V response spectral ratio displays, for the weak seismic events, amplification at the predominant frequencies of 14 Hz (f_2), 3 Hz (f_1) and 0.6 Hz (f_0), which may be related to the depth of impedance contrasts between the geotechnical units, as depicted from the 2D subsurface model.

- The predominant frequencies f_2 and f_1 shift to lower values when the magnitude level increases.

The dynamic modeling based on the Martin–Finn–Seed's pore-water pressure model reveals the following:

- The peak vertical acceleration computed at seismic station AQU noticeably underestimates the observed value, which suggests that some near-source effects probably amplified the vertical component of ground motion in historical L'Aquila City center during the main shock.

- The H/V response spectral ratio reveals that only the predominant frequency f_1 shifts to lower values (from 3 Hz to 1-2 Hz) when the earthquake magnitude increases, whereas the frequency f_2 stays still at 10-11 Hz. This suggests that nonlinear effective stress behavior of soil affects the variations of frequency f_1 and the observed decrease of frequency f_2 (from 14 to 9 Hz) may be associated to a peculiar wave-field incidence angle.

- The nonlinear seismic response of subsurface at station AQU may be related to the generation of excess pore-water pressure in the Quaternary fluvial-lacustrine sequence, which ranges between 50 and 350 kPa during the 2009 M_w 6.1 L'Aquila main shock.

- The seismic motion amplification, which occurs at the shallow colluvium, at the top and base of the fluvial-lacustrine sequence, is consistent with the observed predominant frequencies.

Most of the previous studies, which aimed to evaluate the seismic response in the Aterno River Valley, found evidence of frequency shift due to nonlinear soil behavior, but they usually never considered seismic station AQU in historical Downtown L'Aquila. Comparing the H/V response spectral ratios from ground motion records with those from finite element nonlinear dynamic model, this study correctly estimates the predominant frequencies, which are related to the position of computed seismic motion amplification, as well as the main frequency shift that is associated to the nonlinear behavior of sediments underlying L'Aquila City center. Furthermore, this research reconstructs a reliable model of the subsurface that is based on detailed geological, geotechnical and geophysical data, which may be used by other authors to investigate the factors affecting the seismic ground response in L'Aquila Basin.

References

[1] Amoroso S, DelMonaco F, Di Eusebio F, Monaco P, Taddei B, Tallini M, Totani F, Totani G. Campagna di indagini geologiche, geotecniche e geofisiche per lo studio della risposta sismica locale della città dell'Aquila: lastratigrafia dei sondaggi (giugno–agosto2010). Report CERFI Sn.1;2010. ⟨http//www.cerfis.it/en/download/cat_view/67-pubblicazioni-cerfis/68-reports⟩.

[2] Blumetti AM, Cavinato GP, Tallini M. Evoluzione Plio-Quaternaria della conca di L'Aquila-Scoppito: Studio Preliminare. Il Quaternario 1996;9:281–6.

[3] Bonilla LF, Archuleta RJ, Lavallée D. Histeretic and dilatant behavior of cohesionless soil sand their effects on non-linear site response: field data observation and modelling.Bull Seism Soc Am 2005;95:2373–95.

[4] Bozorgnia Y, Campbell KW. The vertical-to-horizontal spectral ratio and tentative procedures for developing simplified V/H and vertical design spectra. J Earthq Eng 2004;8:175–207.

[5] Byrne PM. A cyclic shear-volume coupling and pore pressure model for sand. In: Proceedings of the 2nd international conference on recent advances in geotechnical earthquake and soil dynamics, St. Louis (Missouri);1March1991.p.47–56.

[6] Chioccarelli E, Iervolino I. Near Source seismic demand and pulse-like record: a discussion for L'Aquila earthquake. Earthq Eng Struct Dyn 2010;39:1039–62.

[7] Chopra AK. Dynamics of structures: theory and applications to earthquake engineering. Englewood Cliffs, NJ: PrenticeHall;1995.

[8] Davoodi M, Jafari MK, Hadiani N. Seismic response of embankment dams under near-fault and far-field ground motion excitation. Eng Geol 2013;158:66–76.

[9] De Luca G, Marcucci S, Milana G, Sanò T. Evidence of low-frequency amplification in the city of L'Aquila, central Italy, through a multidisciplinary approach including strong and weak motion data, ambient noise and numerical modeling. Bull Seism Soc Am 2005;95:1469–81.

[10] De Martin F, Kawase H, Modaressi-Farahmand Razavi A. Nonlinear soil response of a borehole station based on one-dimensional inversion during the 2005 Fukuoka Prefecture western offshore earthquake.Bull Seism Soc Am 2010;100:151–71.

[11] Del Monaco F, Tallini M, De Rose C, Durante F. HVNSR survey in historical downtown L'Aquila (centralItaly): Site resonance properties vs. subsoil model. Eng Geol 2013;158:34–47.

[12] Ditommaso R, Mucciarelli M, Ponzo FC. Analysis of non-stationary structural systems by using a band-variable filter. Bull Earthq Eng 2012;10:895–911.

[13] Di Fiore V, Cavuoto G, Del Monaco F, Mancini M, Caielli G, Cavinato GP, De Franco R, Pelosi N, Rapolla A,Tallini M. Seismic surveys integrated with geological data for in-depth investigation of Mt. Pettino active fault area (Western L'Aquila Basin). Ital J Geosci 2012:131.

[14] Di Capua G, Lanzo G, Pessina V, Peppoloni S, Scasserra G. The recording stations of the Italian strong motion network: Geological information and site classification. Bull Earthq Eng 2011;9:1779–96.

[15] Eskişar T, Kuruoğlu M, Altun S,Özyalın S,Yılmaz HR. Site response of deep alluvial deposits in the northern coast of İzmir Bay(Turkey) and a microzonation study based on geotechnical aspects. Eng Geol 2014;172:95–116.

[16] Field EH, Jacob KH. A comparison and test of various site response estimation techniques, including three that are not reference site dependent. Bull Seism Soc Am1995;85:1127–43.

[17] Gaudiosi I, Del Monaco F, Milana G, Tallini M. Site effects in the Aterno River Valley (L'Aquila,Italy): comparison between empirical and 2D numerical modelling starting from April 6th 2009 Mw 6.3 earthquake. Bull Earthq Eng 2014;12:697–716.

[18] GE.MI.NA. Ligniti e torbe dell'Italia continentale. In: ILTE (Ed.), Torino IP;1963.

[19] Ghofrani HM, Atkinson G, Goda K. Implications of the 2011 M9.0 Tohoku Japan earthquake for the treatment of site effects in large earthquakes.Bull Earthq Eng 2013;11:171–203.

[20] Gorini A, Nicoletti M, Marsan P, Bianconi R, De Nardis R, Filippi L, Marcucci S, Palma F, Zambonelli E.The Italian strong motion network. Bull Earthq Eng 2010;8:1075–90.

[21] Groholski DR, HashashYMA, Matasovic N. Learning of pore pressure response and dynamic soil behavior from down hole array measurements. Soil Dyn Earthq Eng 2014;61–62:40–56.

[22] Guo Z, Aydin A, Kuszmaul JS. Microtremor recordings in Northern Mississippi. Eng Geol 2014;179:146–57.

[23] Huang HC, Teng TL. An evaluation on H/V ratio vs. spectral ratio for site response estimation using the1994 Northridge earthquake sequences. Pure Appl Geophys1999;156:631–49.

[24] INGV-CNT Seismic Bulletin. Bollettino Sismico Italiano – Istituto Nazionale di Geofisica e Vulcanologia. Centro Nazionale Terremoti. ⟨http://bollettinosismico.rm.ingv.it/⟩.

[25] Kham M, Semblat J-F, Bouden-Romdhane N. Amplification of seismic ground motion in the Tunis basin: numerical BEM simulations vs experimental evidences. Eng Geol 2013;154:80–6.

[26] Konno K, Ohmachi T. Ground-motion characteristics estimated from spectral ratio between horizontal and vertical components of microtremors. Bull Seism Soc Am1998;88:228–41.

[27] Kramer SL. Geotechnical earthquake engineering.Upper Saddle River, NJ: Prentice Hall;1996.

[28] Lermo J, Chavez Garcia FJ. Site effects evaluation using spectral ratios with only one station. Bull Seism Soc Am1993;83:1574–94.

[29] Lanzo G, Pagliaroli A. Seismic site effects at near-fault strong-motion stations along the Aterno RiverValley during the Mw 6.3 2009 L'Aquila earthquake. Soil Dyn Earthq Eng 2012;40:1–14.

[30] Luzi L, Hailemikael S, Bindi D, Pacor F, Mele F, Sabetta F. ITACA (Italian accelerometric archive): a web portal for the dissemination of Italian strong motion data. Seism Res Lett 2008;79:716–22.

[31] Martin GR, Finn WDL, Seed HB. Fundamentals of liquefaction under cyclic loading. J Geotech Eng Div ASCE, GT5 1975:423–38.

[32] Mayeda K, Malagnini L,Walter WR. A new spectral ratio method using narrow band coda envelopes: evidence for non-self similarity in the Hector Mine sequence. Geophy Res Lett 2007;34:L11303.

[33] NakamuraY. A method for dynamic characteristics estimation of subsurface using microtremor on the ground surface. Q Rept Railw Tech Res Inst 1989;30:1.

[34] Newmark NM. A method of computation for structural dynamics. ASCE J Eng Mech Div 1959;85 No EM3.

[35] Nunziata C, Costanzo MR. Ground motion modeling for site effects at L'Aquila and middle Aterno river valley (central Italy) for the Mw 6.3, 2009 earthquake. Soil Dyn Earthq Eng 2014;61–62:107–23.

[36] Pacor F, Paolucci R, Luzi L, Sabetta S, Spinelli A, Gorini A, Nicoletti M, Marcucci S, Filippi L, Dolce M. Overview of the Italian strong motion database ITACA1.0. Bull Earthq Eng 2011;9:1723–39.

[37] Petitta M, Tallini M. Ground water resources and human impacts in a Quaternary intramontane basin (L'Aquila Plain, Central Italy).Water Int 2003;28:57–69.

[38] Protezione Civile. Nazionale, Università degli Studi dell'Aquila. Project microzonazione sismica L'Aquila (May-December2009). Geol Map 2009. http://www.protezionecivile.gov.it/.

[39] Puglia R, Ditommaso R, Pacor F, MucciarelliM, Luzi L, Bianca M. Frequency variation in site response as observed from strong motion data of the L'Aquila (2009) seismic sequence. Bull Earthq Eng 2011;9:869–92.

[40] QUAKE/W. Geo-slope international Ltd. Engineering book. 3rd ed. 2008 Version 2007.

[41] Ragozzino E. Nonlinear seismic response in the western L'Aquila basin (Italy): numerical FEM simulations vs. ground motion records. Eng Geol 2014;174:46–60.

[42] Sawada Y, Taga M, Watanabe M, Nakamoto T, etal. Applicability of microtremor H/V method for KIK-NET strong motion observation sites and Nobi plain. In: Proceedings of the13th World Conference on Earthquake Engineering 2004,Vancouver (B.C., Canada);1–6 August2004. Paperno. 855.

[43] Sawazaki K, Sato H, Nakahara H, Nishimura T. Temporal change in site response caused by earthquake strong motion as revealed from coda spectral ratio measurement.Geophys Res Lett 2006;33:L21303.

[44] Tanimoto T, Ji C, Archuleta R. Inversion and prediction of ground motion of the 2009 L'Aquila Italy Mw 6.3 Earthquake. USGS; 2011. Award Number G10AP00010 2010/1/1-2011/12/3.

[45] Theodulidis N, Bard PY. Horizontal to vertical spectral ratio and geological conditions: an analysis of strong motion data from Greece and Taiwan (SMART-I). Soil Dyn Earthq Eng 1995;14:177–97.

[46] Theodulidis N, Bard PY, Archuleta R, Bouchon M. Horizontal-to-vertical spectral ratio and geological conditions: the case of Garner Valley downhole array in Southern California. Bull Seism Soc Am 1996;86:306–19.

[47]Vezzani L, Ghisetti F. Carta geologica dell'Abruzzo (Scala1:100.000). S.EL.CA., Firenze; 1998.

[48] Working Group MS–AQ.Microzonazione sismicaper la ricostruzione dell'area aquilana. L'Aquila: Regione Abruzzo – Dipartimento della Protezione Civile; 2010, 3 and Cd-rom.

[49] Wu C, Peng Z, Ben-Zion Y. Non-linearity and temporal changes of fault zone site response associated with strong ground motion. Geophys J Int 2009;176:265–78.

[50] Wu C, Peng Z, Assimaki D. Temporal changes in site response associated with the strong ground motion of the 2004 Mw 6.6 Mid-Niigata earthquake sequences in Japan. Bull Seism Soc Am 2009;99:3487–95.

[51]Wu G.Volume change and residual pore water pressure of saturated granular soils to blast loads. A research report submitted to Natural Sciences and Engineering Research Council of Canada;1996.

[52] Zambonelli E, DeNardis R, Filippi L, Nicoletti M, Dolce M. Performance of the Italian strong motion network during the 2009, L'Aquila seismic sequence (central Italy). Bull Earthq Eng 2010;9:39–65.

YOUR KNOWLEDGE HAS VALUE

- We will publish your bachelor's and
  master's thesis, essays and papers

- Your own eBook and book -
  sold worldwide in all relevant shops

- Earn money with each sale

Upload your text at www.GRIN.com
and publish for free